GORILLA

IN THE

COCKPIT

BREAKING THE HIDDEN PATTERNS
OF PROJECT FAILURE
AND THE SYSTEM FOR SUCCESS

VIP VYAS & DR. THOMAS D. ZWEIFEL

Printed Worldwide
First Printing 2022
First Edition 2022

ISBN   979-8-35-798839-3 — paperback
ISBN   979-8-35-907922-8 — hardcover

GORILLA
IN THE
COCKPIT

PRAISE FOR GORILLA IN THE COCKPIT

The #1 derailer of major and megaprojects, and indeed all change initiatives, is the human factor. Whether you lead a company, a project or your life, *Gorilla in the Cockpit* is a smart investment.
—**Scott A. Snook, Professor, Harvard Business School**

I wish I had had your tools 35 years ago when I was starting out.
—**Werner Brandmayr, Chairman, ConocoPhillips Europe**

Gorilla in the Cockpit is a rare look behind the scenes of megaprojects by two seasoned experts. If you want to understand what really goes on in big projects, why they fail so often and what it takes for them to succeed, this book is a must-read.
—**Bent Flyvbjerg, professor at Oxford and Copenhagen, principal author of** *How Big Things Get Done* **and** *Megaprojects and Risk.*

Highly valuable for our executive board and 150 direct reports in transforming the mindset, communication, and strategic outlook in a way that led to real-world results.
—**Wolfgang Pitz, CEO, SpaceTech; ex-Space Program Manager, Airbus Defense & Space**

Gorilla in the Cockpit demystifies organizational failure. The book shows the tremendous impact of the Black Box effects on the oversight of massive complex programs.
—**Stanislav Sheknia, Senior Affiliate Professor, INSEAD**

Why do mega-projects fail so often, and at such great expense? In this highly readable and authoritative guide, Vip Vyas and Thomas Zweifel home in on the 'black box' of unstated assumptions and biases of those involved, that stymie effective decision making. If you are involved in project management -at whatever scale- they offer practical solutions to your thorniest problems.
—**Julian Birkinshaw, Vice Dean and Professor of Strategy & Entrepreneurship, London Business School**

Gorilla in the Cockpit reveals why so many megaprojects exceed mega-budgets and don't live up to their mega-expectations. The insights in this book are not limited to physical projects. They also apply to major changes in organization strategy and culture. This timely work is relevant and a must-read for all business leaders who share responsibility to deliver mega-results.
—**Ron Kaufman, New York Times bestselling author of** *Uplifting Service*

Hugely enjoyable! True leadership thinking for the 21st century. A myriad of disciplines blended in a practical way for those dealing with the complexity of mega-initiatives.
—**Greg Bernarda, Co-author of** *Value Proposition Design*, *Strategyzer* **Series**

Though mega-projects represent the greatest risk investors, businesses, or the public, will ever take, few consultants have attempted a comprehensive and practical look at what can be done to improve RoI and reduce volatility. Vyas and Zweifel have done that. With an eye on academic research, and decades of field experience, they have produced a critical synthesis of an ignored topic
—**Paul Gibbons, Professor, Best Selling Author of The Science of Organizational Change, Leadership and Culture Partner, IBM Consulting**

We opened the Black Box and addressed the ugly hidden patterns running one of our $700 million projects. This was the first step in creating a wholesale turnaround of performance.
—**Darrel Kingan, ex-Deputy Director, Capital Works, Hong Kong Airport Authority**

Gorilla in the Cockpit provides a high-octane, highly implementable approach to shifting gears on major and megaprojects. I would fully recommend the underlying methodology from the direct experience of working with the authors in shifting the direction of past projects and creating successful outcomes.
—**Penny Hubbard-Brown, ex-Country Manager MACE, Campus Director, HKUST**

Wish I would have had *Gorilla in the Cockpit* 20 years ago while serving on the team that built the Army's transformational plan for the Chief of Staff of the US Army. Vip and Thomas have added deep clarity enabling a better understanding of the critical leadership literacy of project management. This is the "new doctrine"; get it on your bookshelf!
—**Joe LeBoeuf, PhD, Professor of the Practice Emeritus, Fuqua School of Business, Duke University; former Academy Professor, US Military Academy at West Point**

Brilliant. The level of factual detail is astonishing and the conclusions utterly convincing. Admire the clarity of the writing.
—**Michael Gates, Associate Fellow, Säid Business School, University of Oxford**

Finally a book that sheds light on people power in projects—especially on the invisible and extremely important influence of neuroscience, beliefs, and bias on performance.
—**Markus Hotz, Chairman, Insights Schweiz**

Having worked for 35 years on mega-capital projects in over 60 countries, I learned first-hand about the power of the "black box" that splits project failures (unfortunately the norm) from successes. Zweifel and Vyas provide anyone involved in projects big or small with a clear, understandable, and accessible analysis and roadmap–though it requires leadership courage to implement. My experience on over $1 trillion in capital projects proves that their assertions hit the mark. Read this, do what they say, and lead your project to success in all measures: cost, speed, quality, and safety.
—**Jay Greenspan, Founder, JMJ Associates**

The book I was waiting for! All too often I have seen great projects fail despite green lights on all dashboards. The problem is that most managers see only technical, financial, and strategic indicators—but the factors that decide on failure or success of a project are the soft ones. With *"Gorilla in the Cockpit,"* Zweifel and Vyas give access to a sense of power, confidence and stoic calm that comes from flying in a fully functioning cockpit and knowing which levers to push when a storm comes in.
—**Frederic Mueller, ex General Manager, ABB Switzerland; CFO, Wandfluh AG**

As chairman of a large insurance company and as a doctor focusing on patient-centered medicine, I often deal with megaprojects and large-scale change, from drug development to process integration. *Gorilla in the Cockpit* hits the nail on the head: Projects don't fail—people fail. If we are to succeed with projects of any size, we must find a way to master the human component. Zweifel and Vyas go to the source of why 65% of projects crash, destroying millions of lives and billions in shareholder value—and they offer a systematic methodology and case studies to show us how to fix the plane mid-air. An important book, a pleasure to read, and a significant return on investment.
—**Prof. Dr. Thomas D. Szucs, Chairman, Helsana**

Gorilla in the Cockpit is a must-read for project managers—and/or anyone who wants to get things done through other people. Self-management is already demanding; team management raises the bar pretty quickly; company and project management compounds the challenge exponentially with the number of people involved. Why? The more human beings participate in your project, the more human flaws and bias you need to align to achieve your goals or to simply avoid complete failure—since most projects fail. "Gorilla in the Cockpit" provides you with systematic tools to master your projects and unleash the full potential of its human factor to achieve your goals.
—**Philippe Baeriswyl, Executive Director, Eiriz Réalisations et Immobilier SA**

What causes large projects to fail? Søren Kierkegaard said: Life can only be understood backwards; but it must be lived forwards. A project has a beginning and an end. In order to measure it, you need a goal, a deadline, and a cost. But projects fail not because of the measurable "white box" goals, they fail because of the "black box" of the stakeholders. Vip and Thomas describe masterfully how to deal with these. They show a solution, the "project flight path." I can only recommend this rich book. It is based on a wealth of experience and lessons learned. If you want to cross the finish line, you MUST read this book!
—Dr. Alexander Herzog, Head of Staff, PA-FD, Financial Directorate of the Canton of Zurich; Audit Commission, Municipality of Küsnacht

Major failures in business come about because of problems in the human machinery behind the scenes of organizations. It's increasingly difficult for leaders to gain visibility into the engine. With ever more scale and complexity in global business, the machine gets more opaque, not more transparent. *Gorilla in the Cockpit* opens up the black box and provides a revealing study on the machinery and why it goes wrong. It also helps us see how to get it right.
—Chris Howells, Senior Vice President, Teneo

Gorilla in the Cockpit comes just at the right time! In the biotech and pharma industry, every project is by definition a mega-project with long durations and large investments. The sector is in permanent transformation due to a high error rate, Vip Vyas and Thomas Zweifel give us a powerful book: They encourage and empower the individual leader and his or her actions. They help all of us tackle mega-projects without fear, but with self-confidence and humility, to take responsibility, take decisive action, and ultimately achieve success.
—Dr. Erich Greiner, CEO, Cedrus Therapeutics Inc

Valuable book with many pragmatic and good concepts.
—Gaelle Olivier, Chief Operation Officer, Société Générale

A unique book with no substitute. Vip and Thomas convincingly reveal the Black Box's powerful impact on organizations. The approach made a big difference on one of the world's largest Offshore Wind Farm projects we delivered.
—Luc Vandenbulcke, CEO DEME Group

An enthralling read. I wish I had read a copy many years ago as I set out in my construction career.
—Simon Buttery, CEO Continental Engineering Corporation

Taking our business to the next level involved breaking out of the Black Box and creating a safe psychological space to make our employees feel comfortable speaking up and contributing. *Gorilla in the Cockpit* provides a strong compass for leaders to focus their attention on the areas that make the biggest difference in making shifts in thinking, behaviors, and culture.
—Ian Edwards, President and CEO at SNC-Lavalin

Kudos to Vip and Thomas in crafting an interesting and highly readable book underpinned with well-known case studies and anecdotes on infamous failures and successes of megaprojects. While most publications related to project management tends to be stereotypical and unexciting, the authors have been able to cleverly articulate their findings on the causes of megaproject failures with analogies borrowed from the aviation industry. Remedies to avoid failures through their lenses are provided with even a checklist to guide readers in mitigating disastrous consequences. I do recommend "Gorilla in the Cockpit" as a leisurely read to understand the impact of the human mindset, behavior, culture, and neuroscience on a project. The revelation should be of interest to all those desiring success while undertaking projects of any size.

—Michael K C Yam, President, Chartered Institute of Building (CIOB) & Past President, Real Estate and Housing Developers Association (REHDA) Malaysia

Finally – the ultimate "project cookbook that pulls the hidden agendas, secret motives, and dark derailers from the soft underbelly of projects to the light, so we can put them on the table, examine them, and manage them as we manage all other elements of the project. "Gorilla in the Cockpit" is refreshingly frank, open, and transparent. It gives you the recipe for managing projects with realism and success -without walking over dead bodies – by putting human agents at the center. The case studies are priceless, offering best practices from the rare megaprojects that have succeeded. A must-read.

—Willi Helbling, CEO, BPN Business Professionals Network

TABLE OF CONTENTS

Prologue: The Flightdeck Reality

You are sitting in the pilot seat, hands on the control column,

eyes darting across a range of instruments in front of you.

It's a $10 billion project. By far the largest you've ever managed.

All eyes are on you.

The investors are impatient, almost ruthless, and your "crew" is nervous.

You look out the window and see a violent storm system heading your way.

You are peering directly into the belly of your project ecosystem.

The project has barely begun, but you are already behind schedule.

The Immigration Department has been slow processing visas,

and several key team members are not on board yet.

Some of the specialized equipment you need is stuck in customs.

Procurement has started, but contractors and vendors are submitting prices way above your initial estimates.

It looks like hitting the budget is already going to be a struggle.

Last week, two workers got electrocuted and are in critical condition.

The local press, a key stakeholder, has jumped onto the accident,

and the families are seeking colossal compensation.

You haven't had a good night's sleep in three days (or weeks, months?).

You are about to take a long deep relaxing breath,

but a ping from your phone snatches your attention.

It's an SMS. More bad news.

The message says that the two large engineering firms you have contracted are struggling to attract experts.

You nod to yourself and mutter,

"Not surprising. The project is in the middle of nowhere.

and the accommodation, well, it's certainly not the Park Hyatt."

You absorb the information and then remind yourself

to bring it to the next project board meeting in three days.

Welcome to the wild world of major and megaprojects.

WHY THIS BOOK MATTERS

To undertake a project,

as the word's derivation indicates,

means to cast an idea out ahead of oneself

so that it gains autonomy and is fulfilled

not only by the efforts of its originator but,

indeed, independently of him (her) as well.

— Cieslav Milosz

We know that a full 65 percent of all megaprojects fail. Either they go over budget, over time, or both. Or they don't meet their objectives. At a current investment of $20 trillion per year in major projects, this would be like flushing $13 trillion down the drain.[1] That is the number "13" followed by 12 zeros: 13,000,000,000,000. To put the string of zeros in perspective, consider that a trillion dollars would buy you a $6 Starbucks latte every day, for the next 450 million years.[2]

Did you also know that by 2027, an estimated 88 million people will work in project management-related roles?[3] Again, this is a rather large number. What factors are driving this trend? We observe at least four. The first key is the projects and programs needed to combat unprecedented challenges facing humanity—climate change, water shortages, and global food insecurity being just three examples. A second driver is the massive Chinese "Belt & Road" and the U.S. "Build Back Better" initiatives designed to reconfigure, boost and expand the primary, secondary, and tertiary sectors of two gigantic economies battling for geopolitical supremacy. Both have already spurred enormous demand for advanced project leadership. Then there is the impact of the Covid-19 pandemic—governments worldwide are making significant investments to help their sluggish economies rebound. Last but not least is the exciting lure of emerging disruptive technologies—from the Metaverse to the Blockchain, from biotech to med-tech, from fin-tech to AI —where business angels, venture capitalists, family offices, and wealth managers are eager to swoop upon opportunities for exponential returns.

The project volume is there, but so is the risk. And we have a simple choice: We will either keep going as we always have, playing major and megaprojects like a lottery and usually (two out of three projects) failing. Or we will finally get to the bottom of why megaprojects systematically go wrong, so we can course-correct at the root cause level and put in place the critical path to have megaprojects succeed.

Whether you are a client or a project manager, the tools and technology in this book will enable you—assuming you apply them correctly—to (1) design and set a new project on a path to success or (2) turn around your existing project, steer it to success, and in the process save billions of dollars (or whatever your preferred currency).

Specifically, we have written this book for people who have a genuine interest in shaping the future, creating long-term value, de-risking investments, leading change in complex organizations, driving new levels of performance, and making big things happen with greater certainty, more efficiency and lower costs.

The very word "project" comes from the Latin *pro* (ahead, forward) and *iectum* (thrown). As Cieslav Milosz put it in the motto above, "To undertake a project, as the word's derivation indicates, means to cast an idea out ahead of oneself...." Think of a project, big or small, whether it is to run a marathon or build a transnational pipeline, that you have undertaken. What makes any project unique is that it translates your imagination into reality—it is the vehicle by which you forge, fashion, and design the future. Projects are, in essence, the agents of change and transformation.

Understanding "System for Success"

There are piles of books on project management: A recent search yielded over 20,000 search results on Amazon alone. A smaller number of specialized books cover specific project methodologies such as Prince, Agile (including Scrum), Kanban, and others. The bookshelf becomes extremely sparse in providing a ground zero, eye-level view of how large projects function in real-time. This gap stems from the fact that relatively few people have worked on complex megaprojects from initiation to final operations. An even smaller number has the time to document, analyze and articulate their experiences in a meaningful way. And virtually nobody has deep insight into the human dynamics of megaprojects.

This is a significant gap in the field of project management. We aim to bridge that gap based on our over 50+ years of combined experience working on large projects. As will become clear in this book, many projects create an invisible "system of failure" that predictably derails the project and has it spinning out of control. By "System for Success," we don't mean a silver bullet or a complex checklist. The field of project management is already littered with tons of these, and many of them are very useful. Instead, "System for Success" is about creating a project environment, a force field that sets the project up to win, even when the pressure is on, and the circumstances look ugly.

WHO SHOULD READ THIS BOOK

Anytime you write a book, it's imperative to understand the needs of your audience. We have kept the following key stakeholders in mind:

Investors, even the best and brightest among them, might underestimate the complex risks that come with megaprojects. Why would you invest billions in a venture with a 70% chance of failure, have the project fail, and then do it again? The book aims to help investors understand, appreciate, and actively reduce the risks of their investments. Once you know the hidden risk factors, you can improve your due diligence before you sink real money into a project. And if you apply the Flight Path framework correctly, you will make better investment decisions, de-risk your investments, and maximize your returns. (No guarantees, of course.)

Boards, CXOs, and Project Sponsors, despite knowing that the buck "ultimately stops with us," might lack the experience to ask the right questions, especially when it comes to unique and colossal endeavors. The book aims to provide them with the necessary dashboard to watch and prevent potential crash landings, no matter how large the project.

Project Directors are usually technically solid and intimately familiar with the operations of their megaprojects but might get caught up in day-to-day fire-fighting and crisis management. By revealing the invisible drivers shaping current performance, *Gorilla in the Cockpit* aims to help project leaders enhance their total view of the project, show the leverage points for effective interventions, and enable decisive action.

Change Agents are strongly oriented to address the visible factors of project performance (or what we will call the White Box). They might have minimal training (in some cases zero) in tackling the complex dynamics of the hidden performance factors (what we will call the Black Box). Once change agents are conscious of these hidden dynamics and have learned how to deal with them effectively, they can interrupt the vicious cycle in which projects are caught so often and turn their megaproject around.

Delivery Teams might fall into the trap of viewing and tackling project issues with a mechanical, cause-and-effect, linear mindset. This can often solve only a tiny fraction of the challenges at play or solve the wrong problem altogether. The book helps you detect and address the "nudging factors" that can easily knock your projects off track.

Academics & Students both represent the future resource and capability of the economy. Our Flight Path framework aims to give you an exponential leap in understanding complex initiatives and their value realization and open up new frontiers for future research and studies of megaprojects.

Ultimately, *Gorilla in the Cockpit* is for anyone with an ambitious project. Take us as an example: Writing and producing this book was, of course, far from a megaproject (although at times it felt that way). It was a speck of dust in comparison. But while this micro-project was minuscule in scale, it was every bit as complex. We live in two different time zones, Hong Kong and Switzerland. We saw each other face-to-face only once, for a summer dinner at the Baur au Lac hotel in Zurich, during the five years it took to write the book. We come from vastly different cultures. Vip, a British Indian, likes to explore issues by speaking and brainstorming with industry experts and practitioners. He loves talking and can be rather lengthy. Thomas, a Swiss, is detail-oriented to the point of being pedantic and averse to talking unless it's needed for action. This is just one dimension of the culture clash we experienced; we will spare you the others. And the project of this book had all the ingredients of a project of any size: self-leadership, checking our assumptions, building solid relationships, visioning, business model design, planning, procurement, operations, budget, etc.

Or take the project of raising a family: Anybody who has undertaken to have kids knows what a megaproject that is. OK, raising children might not cost $1 billion, more like $1 million-plus. At least not in terms of money. But in terms of opportunity costs, time, emotions, complexity, and sheer nerves, it is just as intense and costly as any megaproject. So, whether or not you are undertaking a megaproject in the strict sense of the word, this book can be your mountain guide in treacherous terrain.

PREFACE (DON'T SKIP)

It ain't what you don't know

that gets you into trouble.

It's what you know for sure

that just ain't so.

— Mark Twain

(Not sure Twain ever actually said this—but as the Italians put it, "Se non è vero, è ben trovato": Even if it's not by Twain, it's a good one.)

Seriously: If you were planning to skip this preface, this is precisely what happens in large-scale projects. People rush to the "real stuff" (the execution) and often ignore the context. So how about investing the 4 minutes it takes to read this preface? We promise it will be worth your while. So, here goes:

In the pre-Corona world, the consensus was that a megaproject like developing and bringing to market a pharma product would take ten years. The idea that we could launch a vaccine within ten months was ludicrous. But we have done it. So yes, we can, but how?

To answer that question, we have to take a step back: As the French put it, *Il faut reculer pour mieux sauter* (You must step back to leap forward better). Megaprojects are almost as old as history. Suppose we rewind the clock by thousands of years. In that case, we see that the mythical Tower of Babel was possibly the first megaproject—and the first megaproject failure, which led to multiple languages, people being spread across the world, things getting lost in translation, misunderstandings, culture clashes, and wars. Later, the Pyramids of Egypt, the 5,000 km long Great Wall of China, and the Roman Colosseum might be deemed successes since they still stand today. Others, like the Limes, the border wall meant to protect the Roman empire from the barbarians, have crumbled over time. Much later came the Eiffel Tower (1889), the Manhattan Project to build the U.S. atomic bomb (1944), and Soviet Russia's Sputnik (1957).

These megaprojects were few and far between. Today, the market for megaprojects—projects that cost more than $1 billion—is massive and growing at breakneck speed. Within the decade, megaprojects like tunnels, bridges, dams, highways, airports, hospitals, skyscrapers, cruise ships, wind farms, offshore oil and gas rigs, aluminum smelters, communications systems, Olympic Games, satellites and aerospace missions, particle accelerators, and entirely new cities will total an estimated 24 percent of world GDP.

That's one-quarter of all transactions worldwide.

The newest megaprojects are in the high-tech field: Artificial Intelligence (AI), Internet of Things (IoT), 5G, Big Data, and Blockchain. These projects are bound to disrupt or transform the very companies that commission them.

Whether or not you (personally) work on a megaproject right now, you are affected by megaprojects. We all are. Just imagine what our world would be like without airports, hospitals, satellites, bridges, freeways, and shopping malls. For better or worse, megaprojects have an outsized impact on all our lives. When they work, they make our day-to-day far more convenient. And when they fail, when they go over budget or over time, they not only become the laughingstock of entire nations, like the Berlin Brandenburg airport that went nine years over time and €2.5 billion (US$2.98 billion) over budget, in Germany of all places, supposedly a world leader in engineering and efficiency. They also cost all of us real money. German taxpayers followed the disaster with emotions ranging from stoic boredom to rage and very dark humor.

And megaprojects are getting bigger on average. Gone are the days when a project costing $10 million, $50 million, or $100 million seemed impressive. Now Azerbaijan's construction of an artificial archipelago, Turkey's urban renewal project in Istanbul, and Saudi Arabia's Masjid Al Haram each exceed $100 billion.

Big infrastructure projects can also be economically transformative. The Panama Canal accounts for a significant share of the country's GDP. Dubai's international airport is the world's busiest, accounting for 21 percent of Dubai's employment and 27 percent of its GDP. And Hong Kong would surely grind to a halt without its clean and speedy subway system, the MTR, which has enabled the densely packed city to build beyond the downtown districts. Indeed, it is almost impossible to think of these places without megaprojects.[4]

McKinsey estimates that the world needs to spend about $57 trillion on infrastructure by 2030 to enable the anticipated levels of GDP growth globally. Of that, about two-thirds will be required in developing markets, where there are rising middle classes, population growth, urbanization, and increased economic growth.

We (Vip Vyas and Thomas D. Zweifel) have been assessing and turning around projects globally, focusing on human performance for over five decades. We have worked with energy companies, power plants, iconic buildings, irrigation schemes, glitzy casinos, airports, automotive plants, satellite builds, life-saving pharmaceuticals, and mining, to name a few. We have been hired to de-risk the building of a 1,100-mile pipeline that cost $5.4 billion and employed 22,000 workers at its peak, impacting project safety, efficiency, and speed of delivery. We have assisted engineering teams, cutting 90 percent of overtime, or 4.5 years, from the time it took to deliver a satellite to the customer. We have coached the roll-out of a new chemical product in over 100 separate jurisdictions.

We have helped build and implement ATM bank machine protocols. We have worked on a joint venture between a cyber security company and a leading steel manufacturer to construct a multi-billion-euro artificial intelligence solution. We have helped test and bring a shop concept to the gas stations across Germany owned by a global tier-one energy company. We have coached project teams to enter the market in India and China and bring a diabetic diagnostic tool to the vast populations of these two most populous nations in the world. And we have helped build Hong Kong's Ocean Park, a HK$5.5 billion (HKD) masterplan. The combined value of the projects they helped speed up or turn around is over $100 billion.

We found over and over again that one thing consistently went wrong in megaprojects. One missing element, a critical ingredient that requires a radical re-think and re-boot of project management. This one thing was not the technical stuff. And it was not what was in the manual or the GANTT chart. It was the stuff in the Background[5], in the shadows. It was in what people did *not* say. Or what they did not even see, though the writing was on the wall. They ignored the warning signs. This book is based on that experience. It will show you the entire dashboard you need to maximize performance reliability.

Escape > Arrive

At the risk of being presumptuous (but after all, you bought this book for a reason), here are one or several things that might be happening in your projects:

- Despite your best efforts, too many projects have failed or are failing.
- The project goals and objectives are poorly defined, and/or the deadlines are unrealistic. Scope creep is

insidious (and creepy). Risk management or assessment is wanting.

- Keeping teams on the same page seems virtually impossible. Project managers lack sufficient team skills. The result: inefficient teamwork.
- Stakeholders mistrust each other. Miscommunications cause conflicts. Contractors are unreliable. There is no genuine partnership.
- There is a culture of "Cover Your Ass" (CYA), pretense, lies, and even cheating.
- There are lawsuits by dissatisfied alliance partners or clients or suppliers.
- The project management software is not quite right. Different stakeholders might even use incompatible systems. Things fall through the cracks.
- It's not you, of course, or at least not you alone. You are merely a cog in a vast wheel with minimal influence on the outcomes. Truth be told, periodically, you feel powerless. You and/or your colleagues live in resignation. "I make zero difference."
- Some team members have gone into internal exile and demonstrate a frightening lack of accountability.
- You and/or your colleagues feel like you are suffering from the Peter Principle: You have been competent and risen to a level of responsibility that outstrips your capacity. It's unlikely you will get a more significant project.
- This significant failure rate has put a lid on the careers of all those involved, including yours. In a toxic environment of blame games, you must constantly be on guard in a minefield of attacks and accusations.

- You are afraid you will fail ("Project Panic"). As we call it, you and/or your colleagues exhibit survival behavior. You dread going to work.

Does any of this sound familiar?

If you didn't like that list that much, how about this? This is what can happen if you apply this book in practice:

- You will be in the cockpit and steer the plane. You will have more power and control.
- Your success rate will go up by X10.
- You will be known in the company and industry as a project guru, admired, and respected by your peers.
- You will have a team actually working together, jointly delivering success.
- You will have peace of mind, ease, and grace.
- And lo and behold, you'll have fun doing all this.

What if there were one thing that if you put in place in your projects helped you get from what you want to leave behind (we call it Escape) to where you want to be (Arrive)? We offer to give you that one thing in this book. This one thing is the missing component that will provide you with access to delivering projects effectively. (You'll still have to do it yourself, though. There are no guarantees, of course.)

When megaproject expert Bent Flyvbjerg looked at what goes wrong with megaprojects, he researched extensively the political, strategic, and operational crash factors, making vital contributions to the field. In his analysis (and theory), he has also directed the industry towards human factors, such as bias. But, as Einstein is supposed to have said, In theory, theory and practice are the same; in practice, they are not. The human element, with all its facets, from blind spots to bias to culture and communication, remains a black box. We wanted to look at projects from a different angle. The angle we look from is our direct experience with projects. In this book, we open the black box. And we aim to make the human component accessible and actionable.

The Book as a (Micro-)Project

We love projects. Wait, that sentence needs rephrasing: We love to see people succeed in their projects.

This book itself is a micro-project. We live on two different continents. Vyas is based in Hong Kong, Zweifel in Zurich. We are separated by six time zones and thousands of kilometers. The Covid-19 pandemic didn't make it easier. We were prevented from ever getting together (the last time we saw each other was in July 2018 at the Zurich Baur-au-Lac Hotel). We had to communicate via Skype and email. We had to align our strategies, preferences, incentives, mindsets, and value systems. We had to walk the talk and practice what we preach. It became a laboratory of project management in action. And whether you're cooking a meal for four people or 4,000 people, it's only a difference of scale.

There are thousands of books on project management: It has become a cottage industry. What's different about this book? Most books or workshops cover the fundamental mechanics of project management. We will not cover these basic tools in this book. Projects are getting bigger, and people-based projects cannot be managed mechanistically. There is no book out there that systematically covers the hidden factors of project management: mindset, bias, and culture. But as Peter Drucker famously observed, Culture eats strategy for breakfast.

To move outside the existing paradigm of project management thinking and practices, we have taken a design-thinking approach to project management by reviewing project management through many distinct disciplines, including Behavioral Economics,

Linguistics, Ontology, Social Psychology, Neuroscience, and Complex Adaptive Systems.

What you are holding in your hands is the product of that synthesis.

Overview

Each chapter in the book allows you to pass one critical milestone on the way from Escape to Arrive.

Chapter 1 tells the dramatic story of not one but two Boeing crashes that dramatically illustrate how megaproject failures are never just technological disasters. You will see the full spectrum of complexities: the interplay of political forces, financial interests, and human dynamics that underpinned the Boeing organization and led to the disasters.

Chapter 2 dives below the surface of existing project management methodologies and asks a tricky question about their effectiveness given the high observed rate of project failures.

Chapter 3 illustrates how modern project management has advanced from its roots at the dawn of civilization, where the ancients-built marvels of such scale and size that many still exist with us today. Despite the staggering accomplishments, the evolution of project management rests on a faulty foundation that crumbles as projects get larger and larger.

Chapter 4 goes to the root causes for why so many megaprojects crash and why they chronically go over time and/or over budget. The radar that project leaders currently use is inadequate, so they don't see the warning signs until too late. Much project management suffers from a fundamental flaw: a mechanistic paradigm. We need a new project paradigm that treats the human being as the vital pilot of change.

Chapter 5 offers a new model and framework for managing major projects from beginning to end, with nothing left out but still streamlined and lean. We call it the Flight Path.

Chapter 6 shows the dramatic impact invisible factors hidden in the Black Box can have on derailing big projects. The chapter draws a critical distinction between the Foreground and the Background of projects.

Chapter 7 demonstrates the destructive impact the Black Box had on important events leading up to the Deepwater Horizon disaster.

Chapter 8 zooms into the Black Box at the microscopic level and takes a deep dive into the world of the brain and neuroscience. We promise not to overwhelm you -; The chapter delves into just enough science for you to apply to your project and/or change initiative, and spot any project killers lurking unseen, invisibly sabotaging your project.

Chapter 9 gives you tools for building your own "project radar," enabling you to see the warning signs and diagnose the current state of your project more accurately.

Chapter 10 applies the model of the Flight Path and its Black Box to the first of four mini-cases, an icon of world architecture, the Sydney Opera House. While the final product is a mesmerizing piece of architecture, its Flight Path was nowhere near as dignified or smooth.

Chapter 11 (not to be confused with bankruptcy under US law, but perhaps aptly named, given its topic) focuses on an entirely different industry for the second mini-case: the defense sector and the highly controversial F-35 Strike Fighter. With program cost blowouts exceeding USD 1.5 trillion, the sheer scale of this program was too vast to ignore.

Chapter 12 shifts focus from project failures to a spectacular success: the Guggenheim Museum in Bilbao. Our third mini-case reveals key success factors that shaped the project's trajectory. As you read this mini-case, inquire into what impact applying these factors could make to your existing or upcoming initiatives.

Chapter 13 provides an example of delivering a "Mission Impossible." The fourth and final mini-case is the colossal Shell-Gas-to-Liquids (GTL) facility - the world's largest plant turning natural gas into cleaner-burning fuels and lubricants. In our mini-case study, we spotlight the depth of commitment, intelligence, and leadership displayed by Shell (and its supply chain) in creating a purpose-driven mindset supported by an array of concrete actions to achieve world-class performance.

Chapter 14 has been designed to elevate your leadership intelligence. Especially when you are being tossed and turned around inside the belly of the project's storm system, right in the "heat of the battle." We provide you with a robust framework of Principles and Practices to enable you to navigate the Flight Path with precision and potency, enhancing your leadership effectiveness along the route.

Chapter 15 weaves together the Flight Path and emerging technologies. The chapter offers a brief tour-de-force of how recent breakthroughs in Internet 3.0, AI, and the metaverse will shape and impact the future of project management.

Chapter 16 is a call to action--actions that build on the book's contents. Expand your organizational capability, hone your diagnostics skills, and impact the flight path of your projects by choosing the most appropriate next step for your business and yourself.

Acknowledgments

A book like *Gorilla in the Cockpit* cannot be produced by the co-authors alone. In many ways, this book is the product of our combined experience assisting CEOs, CXOs, project leaders, and the people we met along the way. Naming them all would require an entire book, so we will acknowledge a few selected individuals.

We thank, above all, our clients at ABB, Airbus, BP, Borealis, Cathay Pacific, Chevron, China Light & Power (CLP), ConocoPhillips, Credit Suisse, Danone, Dreamlab, Faurecia, Fiat, Glencore, Hong Kong Airport Authority, J&J, MACE, Merlin, MTR, Mubadala Petroleum, Ocean Park, Qatargas, RasGas, Sanofi, SC Johnson, Singapore Ports Authority, Sumitomo, Total, UBS, West Kowloon Terminus & M+ Museum, Wynn Macao, and many others who trusted us with their megaprojects and permitted us to do whatever was needed for turnaround.

We thank the hundreds of project managers, thousands of supervisors, and tens of thousands of field personnel who have contributed their expertise in helping us develop the contents of this book.

We thank Bent Flyvbjerg and Alex Budzier for their valuable data analytics and unique contributions to megaproject management; Mark Utting and the team at Turner & Townsend Switzerland, who provided illuminating case studies; Beat Gygi at the Swiss weekly *Weltwoche* and the late Christian Pohnke at Swiss Re, who assisted with analyses on the Boeing 737 Max crashes; Luc Gerardin, Martin Haller, and William X. Meyer for insightful interviews: David Siegel for the list of Fast Projects; Scott Wolf for sharing his wealth of significant projects experience, and continual high-value feedback on shaping (and reshaping) the evolution of this book; Patrick Moore for proofreading and providing constructive feedback on the initial version; and Sarina Mahasiri for providing a her project financing knowledge and direct feedback. Special thanks to the numerous consultants at Shell, both internal and external, including our former colleagues Jay Greenspan, the founder of JMJ, and Rick Bair, JMJ Executive Sponsor, for their input on the Shell Pearl GTL case study. And finally, Simon Moorhouse for invaluable last-minute tweaks.

Our families: Hartly and Freya in Hong Kong; Gabrielle and Hannah (and of course our dog Motek) in Zurich; and Tina Léa in Tel Aviv, for their everlasting support and cheerleading. You give us a future to live for. You are our favorite megaprojects.

Bottom Line

In case you did not heed our suggestion to read the entire preface (we know who you are, you can't hide), here is the bottom line for you. You will love this book if you:

- Want to deliver projects with greater predictability, accuracy, and certainty.
- Are interested in developing an aerial view of how projects fail (or succeed).
- Want a framework that reduces the complexity of the many moving parts and focuses on those issues that will make the most significant impact.
- Want to de-risk your project by asking the right questions.
- Are keen on living a life with less hassle and greater productivity.
- This book will help you:
- Take first steps into a wider world and get a more detailed and more expansive view of what happens on complex projects.

- See the big picture of why projects fail and the patterns of failure.
- Create a deeper understanding of your project.
- Leverage your team's experience, skills, and capabilities.
- Understand what could be driving the behaviors of the various project stakeholders.
- Make the learning from other project failures more real.
- Spotlight the key actions your project should take.

OK, ready? Let's go.

Radar Alert 1

EVENTS ARE NEVER JUST EVENTS.

THEY FORM PART OF A BIGGER PICTURE AND PROCESS.

Chapter 1 > Gorillas in Action: Boeing's Flights of Horror

If you want to be a millionaire,

start with a billion dollars

and launch a new airline.

— Richard Branson

On 10 March 2019, at 8:38 am local time, Flight ET302, an Ethiopian Airlines plane, takes off from Addis Abeba for a two-hour flight to Nairobi.

What happens next is painful to read. It is an inescapable spiral toward death, akin to watching an accident happening in slow motion. The pilots struggled for five minutes with the plane's automated control system.

The Perfect Storm

At 08:38, a sensor on the pilot's side falsely indicates that the plane is close to stalling, triggering MCAS and pushing down the nose of the aircraft.

At 08:39-40, the pilots try to counter this by adjusting the angle of stabilizers on the plane's tail using electrical switches on their control wheels to bring the nose back up.

At 08:40, they disable the electrical system powering the software that pushed the nose down.

At 08:41, The crew attempt to control the stabilizers manually with wheels - something difficult to do while traveling at high speed.

At 08:43, when this doesn't work, the pilots turn the electricity back on and again try to move the stabilizers. However, the automated system engages again, and the plane goes into a dive from which it never recovers.[6] All 167 people on board are killed.

The Impact

Among the victims were 32 Kenyans, 18 Canadians, nine Ethiopians, and eight Americans, plus people from Austria, Belgium, China, Egypt, France, Germany, India, Ireland, Israel, Italy, Morocco, Norway, Poland, Russia, Slovakia, Spain, Sweden, Togo, and the United Kingdom. UN Secretary-General António Guterres described the crash as a "global tragedy." Many passengers were affiliated with the UN or had been on their way to an environmental conference in Nairobi.

Investigators ruled out wrongdoing by the pilots, who had acted flawlessly, and by Ethiopian Airlines, which is seen as Africa's preeminent airline and enjoys a highly professional reputation.

Sadly, the crash in Addis was the second crash in less than half a year. In October 2018, a Lion Air flight crashed in Indonesia, leaving 189 dead.

The U.S. Federal Aviation Administration (FAA), investigating the Ethiopian Airlines crash and the Ethiopian National Transportation Safety Board, found that evidence collected and satellite data showed both flights behaved "very similarly." "The evidence we found on the ground made it even more likely the flight path was very close to Lion Air's," said Dan Elwell, acting administrator at the FAA.[7]

Another chilling similarity: Both planes that crashed were Boeing 737 MAX.

What was the source of the crash? What went wrong?

"Software Issue": Faking Cause & Effect

Boeing claimed the crash factor was a software issue and announced it would upgrade the flight simulator software. But the company's assertion is disputed. Trevor Sumner, CEO of Perch Experience, whose brother-in-law Dave Kammeyer is both a pilot and software engineer, took to Twitter to argue that Boeing's "software upgrade" was a farce.

"Some people are calling the 737MAX tragedies a #software failure. Here's my response: It's not a software problem.

- It was an *economic problem* that the 737 engines used too much fuel, so they decided to install more efficient engines with bigger fans and make the 737MAX.
- This led to an *airframe problem*. They wanted to use the legacy 737 airframes for economic reasons but needed more ground clearance with bigger engines. The 737 design can't be practically modified to have taller main landing gear. The solution was to mount them higher & more forward.
- This led to an *aerodynamic problem*. The airframe with the engines mounted differently did not have adequately stable handling at high AoA to be certifiable. Boeing decided to create the MCAS system to electronically correct the aircraft's handling deficiencies.

- During the development of the MCAS, there was a *systems engineering problem*. Boeing wanted the most straightforward possible fix that fit their existing systems architecture so that it required minimal engineering rework and minimal new training for pilots and maintenance crews. (...)
- On both ill-fated flights, there was a *sensor problem*. The AoA vane on the 737MAX appeared not to be very reliable and gave wildly wrong readings.
- On Lion Air this was compounded by a *maintenance practices problem*. The previous crew had experienced the same problem and didn't record the situation in the maintenance logbook.
- This was compounded by a *pilot training problem*. On Lion Air, pilots were never even told about the MCAS, and by the time of the Ethiopian flight, there was an emergency AD issued, but no one had done sim training on this failure.
- This was further compounded by an additional *economic problem*. Boeing sells an option package that includes an extra AoA van, and an AoA Disagree light, which lets pilots know that this problem is happening. Both 737MAXes that crashed were delivered without this option. No 737MAX with this option has ever crashed.
- All of this was compounded by a *pilot expertise problem*. If the pilots had correctly and quickly identified the problem and ran the stab trim runaway checklist, they would not have crashed.
- Nowhere in here is there a software problem. The computers & software performed their jobs according to spec without error. The specification was just shitty. The quickest way for Boeing to solve this mess is to call

up the software guys to come up with another band-aid. (...)

- When the software band-aid comes off in a 500mph wind, it's tempting just to blame the band-aid."[8]

Up to the second accident, the 737 MAX had been Boeing's fastest-selling plane in the company's history. More than 4,500 planes had been ordered by 100 different operators worldwide.

Now, scores of airlines canceled their orders. Boeing's stock value fell dramatically. One year later, Boeing asked a representative of the crash victims' families if it would be appropriate for Boeing's CEO Dennis A. Muilenburg to attend the memorial.

The response was swift. "He is not welcome here," said Zipporah Kuria, whose father, Joseph Waithaka, was killed in the Ethiopian Airlines crash. "Whenever his name is said, people's eyes are flooded with tears."[9]

Telling The Truth: Speed & Profits Trump People and Safety

Boeing fired Muilenburg after the 737 MAX calamity. But that action did little to address the source of the crashes. The writing had been on the wall for a long time. Internal documents on the 737 MAX Boeing released in January 2020 are full of late-night trash talk between two Boeing pilots who mocked federal regulators, airline officials, and suppliers and described their colleagues as "idiots," "clowns," or "monkeys."

Many of the messages are from then-737 chief technical pilot Mark Forkner, including some late-night instant message exchanges with his deputy, Patrik Gustavsson.

In one exchange, with Forkner sometimes drinking Grey Goose vodka — "I just like airplanes, football, chicks, and vodka, not in that order," he wrote — and Gustavsson preferring Bowmore Scotch, both talk loosely about their bosses and everyone else they have to deal with in varying derogatory ways.

One pilot who presented to FAA staff mocks the agency's lack of technical knowledge: "It was like dogs watching TV." In another message, the 737 MAX is described as "designed by clowns, who are in turn supervised by monkeys."

"Would you put your family on a MAX-simulator trained aircraft?" one pilot asks, then answers himself: "I wouldn't." His colleague agreed.

India's air safety authority, the Directorate General of Civil Aviation (DGCA), is "apparently even stupider" than another unnamed foreign regulator. And one pilot notes, about dealings with the FAA, "I still haven't been forgiven by God for the covering up I did last year." [10]

"These revelations sicken me," said Michael Stumo, father of 24-year-old Samya Stumo of Massachusetts, United States, who died in the Ethiopian Airlines flight. "The culture at Boeing has eroded horribly," he added. "My daughter is dead as a result." [11]

Of course, the lead pilot's lawyer dismissed the more memorable quotes as bravado, nothing more than some hard-charging guys blowing off steam after work. And Boeing disowned the communications, blaming them on a few rogue employees.

But other, more sober and more damning internal emails reveal that the pilots were under intense pressure from the MAX program leadership. They suggest a troubling culture that put speed above safety. And they point to severe problems with how the MAX was developed and certified.

Robert Clifford, the lead lawyer for the Ethiopian Airlines victims, said the documents will "be used by the families of the victims to show a jury that Boeing was reckless and put profits before safety."

Members of the U.S. House of Representatives were particularly incensed by one document showing that to avoid any need for additional pilot training, Boeing downplayed to the FAA the significance of the new flight control software on the MAX — known as the Maneuvering Characteristics Augmentation System (MCAS) — that was implicated in the two crash flights.

House Transportation and Infrastructure Committee vice-chair Rep. Rick Larsen, D- Everett, said these "efforts to characterize the MCAS software as seemingly inconsequential were a serious mistake."

And that was not the end of it. When Indonesian carrier Lion Air in 2017 asked for simulator training for its pilots, apparently at the suggestion of that country's regulator, also known as DGCA, Forkner scrambled to convince the airline that it shouldn't do so.

He approached DGCA and argued that other regulators didn't require sim training, so why should Indonesia.

This manipulation by Boeing of both its airline customer and a foreign regulator looks damning in hindsight, especially when the first crash was a Lion Air jet. Simulator training might have gone some way to compensate for the over-reliance on cockpit automation and pilots' lack of manual flying experience at some low-cost carriers overseas. This emerged as an issue after the two crashes.

Boeing conceded as much in January 2020 when it reversed course and recommended simulator training for all pilots before the MAX returned to service.

No End in Sight

Alas, that was not to be the end of the Boeing drama. A string of failures hampered Boeing 777 aircraft too. In December 2020, an engine failed during a Japan Airlines flight bound for Tokyo. Japan's Transport Safety Board said a fan blade that broke off from the engine showed signs of metal fatigue. Another blade was broken roughly in half.

And on a weekend in February 2021, the engine powering a United Airlines Holdings Inc. flight broke apart over a town near Denver.

In another incident on the very same day, a Boeing plane dropped engine parts after a midair explosion over the Netherlands. Longtail Aviation Flight 5504, a cargo plane, scattered small metal parts over Meerssen, causing damage and injuring a woman shortly after takeoff.

There were similarities between the incident in Japan and the one in Colorado. All three involved Pratt & Whitney engines. These incidents prompted fresh scrutiny by US, Japanese and Dutch regulators that is still ongoing as of this writing.

The Cover-Up Exposed

"Everybody I talked to at Boeing," whistle-blower "Swampy" told *The New York Times*, "is embarrassed to work there most of the time." After a sigh, he added: "It's just, 'Let's go home.'"[12]

There was pressure on the factory floor that was creating a litany of other problems, for example, debris from construction on Boeing's Dreamliner planes at the Charleston, South Carolina factory. Swampy's job was to inspect airplanes to ensure nothing was left behind inside the aircraft. "So I was called out to an airplane to look at an issue," Swampy said. "And that's when we discovered all this debris, these 3-inch-long titanium slivers laying around. It's just debris everywhere." He discovers that these metal slivers are hanging over the wires that control the plane during his inspection "The risk here is that these metal slivers will migrate into power panels, any kind of power, or any kind of electronic equipment, and short it out and cause a fire. And if it's at 40,000 feet, that's a problem."

Swampy took photos and brought the issue to his manager. In response, "I was removed from it." The manager took him off that plane and gave it to someone else to inspect. Did that plane ever get cleaned? No. "It was delivered without being cleaned." And it was not just metal shavings. Multiple whistle-blowers talked about nuts, bolts, fasteners, rags, bubblewrap, trash, tools, chewing gum, and even a ladder: A ton of stuff was left in the bowels of these aircraft.

"There's a lot of pressure to meet schedule," Swampy said. Managers get judged by their superiors based on the number of jobs they complete on an hourly basis. "And it's held against them if they create defects. So, you know, there's an incentive not to report your defect that you created because it's gonna be held against you."

Boeing vehemently denied that they put speed above safety. And to be fair, no Dreamliner has ever crashed. But the question remains, Has the company put profits above safety? Has it put its customers, and ultimately all passengers in its aircraft, in harm's way?

Making the Complexity Visible

The Boeing tragedy we have outlined is an abridged version of the volumes of in-depth crash investigation paperwork that got produced.

When mega-projects go wrong, the underlying hidden structure of the failure can be dispersed and distributed across many files making the overall comprehension of project failure foggy and obscure.

As authors, we intend to help project practitioners and those interested in projects cut through the volume of complexity.

What if we could see the entire story unfolding on a single page? What if we could reduce volume into clarity and insight. Shifting from laundry lists of risks into focussed performance attention?

Dashboard

The harrowing story of the Boeing 737MAX and its life-and-death consequences, but particularly the psychology of coverups, gave us the title of the book: *Gorilla in the Cockpit.* All too often, the pilots in the cockpits of megaprojects do not have the right dashboard to steer the plane to its destination—and to tell you the truth, we have often felt like gorillas ourselves. Sure, the consequences are not always death and destruction as in the case of the Max 37 (although at times they are, as the cases in healthcare and life sciences below demonstrate). But when project pilots don't have their hands on the engine and instruments, the real-world consequences are always dire. The *Gorilla in the Cockpit* is a metaphor for what happens in large-scale projects. They go from A to B and get into trouble along the way. And they never quite make it to their destination.

But not all of them. To fight the coronavirus, China built a 1,000-bed hospital in ten days. It looks like a miracle, but it's not mysterious. Again and again, whether in pandemics or in war, humans can bring urgency and commit themselves to breakthroughs. If you master the human element, you can go way beyond traditional project implementation and produce performance breakthroughs.

Why are so many projects going so wrong? What's missing in megaprojects? "Everyone assumes it's the technical skills," said Mark Utting, country manager of Turner & Townsend Switzerland, that has helped manage over 200 projects, many of them megaprojects, on every continent except Antarctica. "It's not. Megaprojects have super capable people technically. What's missing is an appreciation of the softer skills you need to make a team function."

Utting distinguishes between personality types that all have to collaborate on a megaproject, like it or not: "You have architects that are usually inventive, creative, inspirational types. You have contractors who just want to get stuff done, saying, 'Don't bore me with details and art.' You have technical engineers who tend to want to solve problems and are intensely detail-focused. And of course, you have the client who might be authoritarian, who has the power, who has the money." The tension and friction are built right into the structure.

Luc Gerardin, an innovation and transformation consultant, has accompanied many megaprojects, including several in international shipping. "I worked on a project with Mersk through IBM," Gerardin said. "We had to build consortiums of stakeholders who can also be competitors, frenemies. Each stakeholder has different clients; some are even clients of others. If they are not aligned, if they don't check their mindsets, they fail."

Project manager William Meyer agreed. "You have to do your homework. Who is the financier? Is it private or public, or mixed-form? In the UK, for example, it's mixed-form: private financing, but for the state. You have stakeholders, banks, customers, government, and consultants. For example, the banks can withhold payments if milestones are not met, or quality is not up to their expectations."

Meyer learned this the hard way. He oversaw projects at ABB, Hitachi Zosen, Schindler, and other multinational capital development firms. "I saw everything from big to small, from sale to handover." His experience taught him that megaprojects never happen in a vacuum; you always have to be aware of the contextual factors that can impact performance. "These are factors we have no influence over. We did a project in the UK, an incinerator. Just from the legal system, we had constraints: They have case law."

The keyword in Meyer's statement is *contextual factors*. Why is it that so many project managers don't see these things? To get a good answer to that question, we have to delve into a micro-history of project management and look at several misconceptions that have plagued the field for far too long.

Radar Alert 2

IF FAILURE IS THE NORM,

WHERE DO WE FIND THE SEEDS OF SUCCESS?

Chapter 2 > Project Management: The Ugly Truth

Every successful person has had failures,

but repeated failure

is no guarantee of eventual success.

— Eric Hoffer

Pick up any standard textbook on project management, and you will encounter a fundamental concept of project management: the Project Life Cycle. The model lays out a project visually as a distinct set of phases from ideation to completion. Project Life Cycles are helpful as they outline the order of actions and activities needed to deliver the project effectively and efficiently. Table 2.1 shows the "Waterfall" models proposed by the Project Management Institute (PMI) and the Association for Project Management (APM).

Association for Project Management (APM)	Project Management Institute (PMI)
1. Concept	1. Initiate
2. Definition	2. Plan
3. Implementation	3. Execute
4. Handover & Close	4. Monitor & Control
5. Operations	5. Close

Table 2.1: Project Life Cycles by APM and PMI

A quick scan of the table shows the logical structure of both models—and how similar they are. In theory, both processes appear neat, linear, and straightforward. You know this: All projects start as an idea that is further refined and defined during the Concept or Initiation phase. Progress between the stages typically involves oversight and approval by senior leadership. The first two phases consist of reviewing the credibility of the business case (the commercial and social value of the project), the quality of the estimates being used, project financing, stakeholder considerations, and the profile of the execution team. In megaprojects, successful passage through Phases 1 and 2 typically involves billions of dollars in financial commitments.

The projects then move into the Execution or Implementation phase. A project is considered complete once it has passed from implementation or execution through commissioning and is finally handed over to the client. At first glance, it may be easy to think: "All this seems straightforward. What's so complicated?"

Speaking of complicated: In 2015, Vyas spoke at the Australian Defense Conference on "Transforming Project Performance." A debate ensued among the project professionals about what makes a project "complex." Vyas said, "You can take the simplest of projects, sprinkle a group of people onto it, and boom—you have a complex or complicated project." There was laughter, followed by nods of recognition. But what made people laugh is not all that amusing.

Back to the frameworks. The PMI, APM, Prince, Agile, and other project methodologies are the unquestioned backbones of change delivery. The question is, How effective have they been?

Empirical Evidence—Chronic Cost Overruns

Megaproject expert Bent Flyvbjerg has built a comprehensive database of megaprojects. His statistical analysis gives a rather unflattering picture of our ability to deliver upon complex big-ticket endeavors. The numbers alone make for worrisome reading (see Table 2.2.)

These findings have been echoed by a host of other industry and academic studies, resulting in the rule of thumb that 60-80% of corporate transformation programs fail.

Table 2.2 shows how project fiascos happen across a wide range of sectors.

Project Cost Overrun (%)
Suez Canal, Egypt 1,900
Scottish Parliament Building, Scotland 1,600
Sydney Opera House, Australia 1,400
Montreal Summer Olympics, Canada 1,300
Concorde supersonic aeroplane, UK, France 1,100
Troy and Greenfield railroad, USA 900
Excalibur Smart Projectile, USA, Sweden 650
Canadian Firearms Registry, Canada 590
Lake Placid Winter Olympics, USA 560
Medicare transaction system, USA 560
National Health Service IT system, UK 550
Bank of Norway headquarters, Norway 440
Furka base tunnel, Switzerland 300
Verrazano Narrow bridge, USA 280
Boston's Big Dig artery/tunnel project, USA 220
Denver international airport, USA 200
Panama canal, Panama 200
Minneapolis Hiawatha light rail line, USA 190
Humber bridge, UK 180
Dublin Port tunnel, Ireland 160
Montreal metro Laval extension, Canada 160
Copenhagen metro, Denmark 150
Boston-New York-Washington railway, USA 130
Great Belt rail tunnel, Denmark 120
London Limehouse road tunnel, UK 110
Brooklyn bridge, USA 100
Shinkansen Joetsu high-speed rail line, Japan 100
Channel tunnel, UK, France 80
Karlsruhe-Bretten light rail, Germany 80

London Jubilee Line extension, UK 80	
Bangkok metro, Thailand 70	
Mexico City metroline, Mexico 60	
High-speed Rail Line South, The Netherlands 60	
Great Belt east bridge, Denmark 50	

Table 2.2: Large-scale projects and cost overruns (Flyvbjerg 2014) [13]

The Inconvenient Numbers

Take the Olympic Games, for example. Held every four years—at least before the Covid-19 pandemic—the Olympics is considered the world's premier international sports event, with over 200 nations competing. Almost all Olympics hosts have a long history of massive commercial losses. The notable exception was Los Angeles, which turned a $200 million profit in 1984, the only location to have produced a return on investment since 1932—when the host had been Los Angeles too. In comparison, Beijing in 2008 generated $3.6 billion in revenues but $40 billion in costs. London in 2012 performed not much better, with $5.2 billion in revenue and $18 billion in expenses. Despite this dismal track record, nations fight tooth and nail for the prestige of hosting the next Games. It is a fascinating example of distorted thinking and optimism bias—factors that are not routinely (or entirely) incorporated into current project management methodologies. Much more about that below.

Flyvbjerg and his colleague Alexander Budzier have compared the spiraling cost overruns of hosting the Olympics, and the limited long-term returns, to the catastrophic impact of wars, pandemics, and other natural disasters such as earthquakes, on national economies. Table 2.2 shows: The pattern of failure reaches far beyond the Olympics. Less prominent projects suffer in similar ways. And since they are not in the fish tank of public scrutiny, they often fare even worse.

Failures in Plain View

The higher the ape climbs,

the more people will see

of his bottom.

— Winston Churchill

Given the sheer magnitude of investments involved, megaprojects often draw public attention. At its best, the media exists not just to serve up the "Three S" (sports, sex, and scandals); it's also a public watchdog for transparency and exposing wrongdoing. Add to that huge budget blowouts and late delivery, and complex projects can quickly become an incubator for exposing corruption, incompetence, or conflicts of interest. Some media coverage might even ridicule officials and project managers or whip up public anger. Since many consumers are addicted to dramatic stories of panic, failure, and public humiliation, the media tend to deliver exactly that to keep viewers glued to their content, not least to maximize advertising revenue. Complex megaprojects provide fertile soil for reporters and news anchors. And if the words "billions" or even "trillions" are in the mix, the story immediately catches the eye, and the reader is drawn to find out more. Look at the following headlines about significant project failures, and there's a fair chance you would be tempted to click on the links:

- "Inside America's Dysfunctional Trillion-Dollar Fighter-Jet Program." [14]
- "Crossrail Delay: Unacceptable Lack of Accountability." [15]
- "Why Sydney's Opera House was the world's biggest planning disaster." [16]

Such often-unwanted public attention can lead to a vicious cycle: It can sow division between various stakeholder groups, confuse or discourage investors, or build anti-coalitions of interest groups seeking to derail the megaproject—especially when a lot of money is involved.

All of this—decades of scientific research, volumes of data, and in-depth analysis, coupled with the widely observed high rates of failure—indicates hidden shortcomings, blockages, and blind spots built into the current project frameworks. If those hidden features were prominent, project stakeholders could tackle them. But they are far from obvious.

We will assert and back our assertion with research and data that the current frameworks and project management approaches are derived from a meta-model of thinking: the model that sowed the seeds of how we run organizations and projects today. To reveal this meta-model, we first need to trace its historical evolution.

Radar Alert 3

CONSIDER: YOU DON'T SEE REALITY AS IT IS.

YOU SEE IT THROUGH TINTED GLASSES.

CHAPTER 3 > PROJECT MANAGEMENT: AN EXTREMELY SHORT HISTORY

It isn't that they cannot find the solution.

It is that they cannot see the problem.

— G.K Chesterton

Albert Einstein is often quoted as saying that we cannot solve a problem at the level of thinking that created the situation in the first place. To develop new sustainable solutions, we must initially check the existing thinking patterns that led to the problem. Once we have done so, we can establish the validity of past thinking and think newly.

The question, "Why do so many projects go wrong so often?" requires us to step back a few centuries and confront the level of thinking and the core assumptions on which traditional project management is based. We have outlined a chronology of critical developments and milestones (see Table 3.1) to aid this process. Starting at the beginning, at the dawn of recorded history, with the construction of the Ancient Wonders like the famed hanging gardens of Babylon, the Great Wall of China, or the Egyptian pyramids, Table 3.1 charts key critical milestones throughout the millennia, leading up to the present flurry of interest in project management with the emergence of powerful technologies such as artificial intelligence (AI), robotics, sensing technologies, and real-time dashboard reporting.

Period	Key Developments & Milestones
2570 BCE - 15th Century	**Pre-History.** The construction of the Ancient Wonders of the World, such as the building and coordination works involved in the construction of jaw-dropping icons such as the Great Pyramids of Giza, The Colossus of Rhodes, the Lighthouse of Alexandria, and the Great Wall of China is generally regarded as humanity's foray into the world of mega-projects.
11th-19th Centuries	**Cathedrals.** The first cathedrals were built already in the 4th century with the Christianization of the Roman Empire before the fall of Rome in 456 CE, but construction of large-scale cathedrals like the one in Salisbury or Notre Dame in Paris that required complex coordination and megaproject management started in the 12th and reached into the 19th century.
19th Century	**Warfare and the General Staff.** Karl von Clausewitz (1780-1831), a Prussian senior army officer and influential military theorist famous for his dictum that "war is merely a continuation of politics with other means" and for his—and the West's—principal treatise on the philosophy of war, *Vom Kriege [On War]*, is still studied at military academies today. Before Clausewitz, armies were led by a country's king or ruler, which made that country vulnerable to extinction, much like in chess: Lose the king, and you lose everything. Armies the world over owe Clausewitz, and a group of enterprising young officers around King Frederick the Great, the idea of the general staff, an essential innovation after the defeat of the Prussian army by Napoleon. The idea was to spread authority across a cadre of officers selected competitively, not on their family lineage. Officers could be trained and exchanged, making the army more like a machine and less vulnerable to defeat. Instead of a single commander at the helm now stood an anonymous "General Headquarters" or G.H.Q. of an elite officer corps, a term still used by most of today's corporations.
1911	**F.W Taylor and Genesis of Scientific Management Thinking.** Born in the Industrial Age, the Principles of Scientific Management represent a machine-oriented view of organizations. In this paradigm of management thinking, the focus was placed on organizing resources, enhancing efficiency, and reducing waste. This was thought to be achievable by systematic management based on rules and procedures to govern human and machine activities.

1917	**The Gantt Chart.** Formal codification of project management's fundamental principles and activities, such as organizing tasks, managing resources, and coordinating across interfaces, started with the Gantt Chart, developed by Henry Gantt, generally considered the forefather of project management. One of its first applications was in the building of the mammoth Hoover Dam in 1931. Considered radical and innovative at the time of its introduction, the Gantt Chart is now one of the most ubiquitous tools in project management used by planners and project teams to monitor and track the progress of their projects. Virtually every certified project manager is familiar with the bar chart for visibly outlining the key workstreams and project phases.
1950s	**Critical Path Method (CPM).** Critical Path is another fundamental concept used in project management. It refers to how long a project will take to complete. More specifically, it is the longest sequence of activities in a project plan that must be completed on time for the entire project to complete by the due date. Introduced by Dupont Corporation, Critical Path was first applied to the complex process of shutting down chemical plants, performing the maintenance works, and then restarting the plants for full production. A critical development at this time was the notion of Single Point Accountability: the idea of having one person accountable for the delivery of a project from concept to completion. **Program Evaluation and Review Technique (PERT).** Initially developed in 1957 for the U.S. Navy Special Projects Office to support the Navy's Polaris nuclear submarine project, PERT has become an industry standard. Like Critical Path, the methodology analyzes the tasks involved in completing a given project, especially the time needed to complete each task. It identifies the minimum time required to complete the total project. PERT incorporates uncertainty by making it possible to schedule a project while not knowing precisely the details and durations of all activities. It is more event-oriented than start- and completion-oriented and is used more in those projects where time, rather than cost, is the primary factor. PERT is applied on large-scale, one-time, complex, non-routine infrastructure and Research and Development projects.
1960s	**Work Breakdown Structure (WBS).** Developed, much like PERT, by the Department of Defense (DOD) during the Polaris mobile submarine-launch ballistic missile project, the Work Breakdown Structure is a reductionist approach to

	breaking a project down into smaller components of deliverable results. It is a hierarchical delivery-oriented structure that enables a project to divide and organize work into manageable sections.
1964	**Configuration Management.** Adopted by NASA, this process documented the functional and physical characteristics of a system. Acting as a baseline, it allowed system changes to be reviewed and documented.
1969 **1970s**	**Project Management Institute (PMI).** Founded in 1969, right after the Grenoble Olympics had used PERT and other project management techniques, PMI was the first formal institute for project managers. **Transaction Cost Economics.** Based on the work of Oliver Williamson that set the transaction as the basic unit of analysis, it became possible to model megaprojects along a continuum ranging from the highly integrated and hierarchical firm with its bureaucracy, all the way to the free market where agents contract with each other but contracts are hard to monitor, police or enforce. In Transaction Cost Economics, megaprojects are located somewhere on that spectrum: As with making a movie, once the megaproject is complete, its structure and the relationships among its stakeholders cease.
1984	**Theory of Constraints (TOC).** In 1984, Eliyahu Goldratt introduced his Theory of Constraints in an unusual fashion: through a novel. In *The Goal*, Goldratt observed that a small number of critical constraints often limit the performance of systems. Built on the principle that "Any chain is as strong as its weakest link," the theory recognizes that the weakest person or part in the system can damage or break the system and jeopardize the outcome. Typical constraints include faulty equipment or the sub-optimal use of equipment; lack of skilled people or counter-productive mindsets; or a written or unwritten policy that stifles performance. The TOC is a focusing process devised by Goldratt to identify those constraints that have the most significant impact on performance. It provides a structural link to the Pareto 80:20 principle, where 80% of effects are derived from 20% of causes. This theory was later expanded and applied to projects, creating the Critical Chain Project Management.
1986	**Scrum.** An agile framework developed for managing software projects, Scrum focuses on team collaboration. It

	takes a complex adaptive approach to engaging product owners, software developers, and end-users to provide "real-time" feedback, ensuring software development (tightly) aligns with customers' existing and emerging needs. Central to the framework is the concept of "Sprints," a series of short timeframes (e.g., four weeks) during which software product features evolve and build the next level of value for the customer.
1989	**Earned Value Management (EVM).** The concept of Earned Value rose to prominence during the 1980s. EVM is a project management approach used to monitor the progress of the project plan via calculating deviations between actual work performed against the work that would have been completed if the project were on track. The calculations establish the amount of value generated to date on the project.
2001	**Agile Manifesto.** In February 2001, 17 software developers calling themselves the Agile Alliance wrote "The Manifesto for Software Development" during a 2-day retreat. Unlike the Scrum Framework, Agile is a philosophy of software development centered on four fundamental values: 1. Individuals and interactions over processes and tools; 2. Working software over comprehensive documentation; 3. Customer collaboration over contract negotiation; and 4. Responding to change over following a plan. The Agile approach to project management was developed in response to the high failure rates of technology projects using the standard step-by-step or phase-by-phase waterfall (cascading) approach. One of the significant issues with the waterfall method was the lack of early user feedback due to the rigidity and linearity of the process. This resulted in the development of large systems not meeting user expectations regarding functionality and usability. Agile's philosophy is to develop "lightweight" software prototypes and test them with users. Using an iterative process, the prototype is rapidly tested with the user base and then further enhanced to ensure product development and customer value are synchronized during the development process.

Now and the Future	The Fourth Industrial Revolution: AI, Blockchain, Metaverse & Co. As of this writing, the phrase coming out of many mouths is "data-driven futures": Artificial Intelligence, Machine Learning, Blockchain, Robotics, the Internet of Things, and various forms of Augmented and Virtual Reality to enhance project planning and monitoring throughout the project life cycle. If this strong (at times even frantic) interest is not just a fad, and it looks like it isn't, these emerging technologies will open new frontiers in megaproject management. Especially the Blockchain, a distributed and 100% transparent ledger, a chain of "smart" (automated) contracts, could act as a "trust engine" among megaproject clients, contractors, and other stakeholders. UBS has already collaborated with IBM to build a use case for an international shipping project (e.g., a shipment of rice from China via the Gulf of Aden to Europe) that drastically reduced transaction costs by eliminating some 35 different bills of lading, and the legal costs of drawing up, monitoring and enforcing contracts among all parties (and hence, sorry, cutting out 35 lawyers and notaries). [17]

Table 3.1: History of Project Management, Key Developments & Milestones

The World Is Not a Clock

Most modern developments in project management can be traced back to the seventeenth century. Isaac Newton gave us the concept of gravity and other fundamental physics laws. His observations regarding the motion of objects were groundbreaking and paved the way for The Enlightenment and a scientific understanding of the world.

Why does the world view of a long-dead philosopher matter? It matters a great deal, as we will see. We have inherited a way of thinking about the world as a clock or a machine. That thinking went into medicine, biology, and other sciences—even economics and politics. The idea was that you could separate things and people and disaggregate them into different components. Then you could see what's wrong, fix it, put it together again, and optimize it.

But projects are far from mechanical; the clock or the machine is the wrong image for a megaproject. The failure of projects has shown that the old approach is bankrupt and obsolete.

While we have the utmost regard for Newton and the enormous value the mechanistic paradigm provides for designing and constructing much of what we have come to take for granted around us (from elegant architecture to medical machines, from urban infrastructure to self-driving cars), the data from the field suggests (see chapter4, Figure 4.1) that the mechanical paradigm is necessary but insufficient for delivering complex projects successfully. The evidence shows the biggest challenges in delivering major and megaprojects are not challenging technical issues but people.

The problem is, if you see everything as a machine, you act accordingly. Devices can be taken apart and dissected so you can analyze them. The assumption is that if you understand each part of the machine, you will understand the whole. Machines are viewed as objects that can be fixed in isolation, not interdependent connections.

This is far from trivial. Since Newton, we have become trapped in viewing reality as something fixed, mechanical, linear, and structural. If you have a problem in your body, you go to fix it structurally. If your foot is injured, you go to a podiatrist, and he will do foot surgery. As a specialist, he will not realize that the source of your foot injury lies in your spine, which comes from a psychosomatic issue due to overwork, which ultimately might originate in your frame of mind. So Newtonian separation and specialization lose sight of the whole system. This goes not only for the human body but for any human endeavor, be it organizations, politics, or projects.

Since the end of the 20th century, scientists in disciplines ranging from physics to biology, from medicine to information technology, have begun to question whether the machine model of the world created most notably by Newton can adequately explain how life (and yes, a megaproject) truly works. Quantum physics, field theory, and chaos theory have shown us that the universe does not work like a giant clock. But our worldview is still shaped by a mechanical model.

By the way, we are not denigrating this mechanical worldview. It indeed has its place, for example, when you build a house, a traffic light system for Hong Kong, or the canalization for Mumbai. As long as no human beings are involved, you can manage the project as you would a clock—a static, linear, and predictable task. But once there are human stakeholders, a Newtonian worldview is woefully inadequate.

This mechanistic worldview has become deeply entrenched, not least in project management. Walk around any project and tune into conversations taking place, and you will hear the machine type thinking right in front of you, embedded in language, in such everyday phrases as:

- *"We need to **lift productivity.**"*
- *"Let's **leverage** our **resources.**"*
- *"Human **resources**" or "human **capital**" or "headcount";*
- *"We need more **bandwidth**";*
- *"We should **break the problem down** into its pieces"; or*
- *"How good are the **control and monitoring** systems currently in place?"*

The Streetlight Effect

The story is well-known: A police officer sees a drunk searching for something under a streetlight and asks what the man has lost. He says he lost his keys. The policeman wants to be helpful, so they look under the streetlight together. After a few minutes, the policeman asks the drunk if he is sure he lost his keys here, and the drunk replies, "No, actually, I lost them in the park." The policeman asks, "So why are you searching here?" The drunk shoots back, "Well, it's way too dark in the park. This is where the light is!"[18]

Have we been looking in the wrong place to discover the keys needed for success in the world of projects?

Have we focused on failure symptoms but been blind to the root causes? Are the keys to understanding project under-performance in the dark, and we have yet to find them? You guessed it: Our answer is an absolute "yes." For decades, and probably since the early days of Taylorism and "Scientific Management," we have looked in the light only. Now, after untold numbers of failed projects and megaprojects, the time has come to look in the dark. Let's arm ourselves with a strong torchlight and see what's really going on.

Radar Alert 4

ARE YOU TRYING TO SOLVE 3-D PROBLEMS

USING A 2-D TOOLKIT?

CHAPTER 4 > FAULTY SENSORS

Don't be trapped by dogma—which is
living with the results of other people's thinking.

— Steve Jobs

Imagine you just walked into a room filled with nervous and worried people. It's a meeting to review the performance of a badly failing project. So bad that the 10-15 people in the room have trouble looking at each other's eyes. The room is filled with shame, embarrassment, and tension. Several participants fear for their jobs; others are seething with barely disguised anger, about to erupt; yet others are rapidly thinking about reasons and excuses to explain away the poor performance.

As project experts, we have attended many such meetings. Sometimes the board or senior management brought us in to prevent the congregation from spiraling down into blame-games and finger-pointing, excuses and shirking responsibility, and the like by those responsible for the poor outcomes.

Once we moved beyond the blame, shame, anger, sadness, and rationalizations, we have worked with organizations to reveal the authentic "deep source" underpinning failures and support our clients to get their hands on the machinery that drives future success. In partnership with the executives, we have exposed those areas where the project has suffered from impaired vision and discovered the pivotal factors that have shaped the results so far.

From our experience, project teams explain away bad performance using a pattern of explanations or rationalizations. This list of Project Crash Factors is a typical output you would likely encounter from a forensic review process of a troubled project:

1	Poorly defined project scope
2	Inadequate risk management
3	Failure to identify key assumptions
4	Project managers who lack experience and training
5	No use of formal methods and strategies
6	Lack of effective communication at all levels
7	Key staff leaving the project and/or company
8	Poor management of expectations
9	Ineffective leadership
10	Lack of detailed documentation
11	Failure to track requirements
12	Failure to track progress
13	Lack of detail in the project plans
14	Inaccurate time and effort estimates
15	Cultural differences in global projects

Table 4.1: The Top 15 Causes of Project Failure [19]

Look at the table above again and ask yourself, "Is there anything people do not influence on the list?" We suspect you will arrive at the same conclusion as we did: There isn't a single item on the list that is not impacted by people's thoughts and actions. In short:

Projects don't fail | people do.

Your project's performance (or lack t
performance of individuals, groups, (
that constitute your project ecosyste
a project that doesn't have the finger
it. It is the people, not the machines, nor the finances, that make (or break) the project.

In other words, a project is a social system where groups of people come together to take collaborative action and collectively solve problems. All of this happens against the backdrop of a commitment to deliver on a set of objectives. And what these people, above all the project leaders, focus on matters enormously. "It's the same process whether you work on the Blockchain, or with diplomats, for example, the Iran deal or negotiations with China," said Luc Gerardin, an innovation and transformation leader who worked extensively with IBM. "Projects fail because project leaders ignore self-awareness or relationship. They focus just on the vision or strategy or action."

Projects Are Human

When we think of organizations and projects from a mechanical perspective, the following happens: First comes the strategy, then the structure, the process, and finally, the execution. We see people as epiphenomenal. As we saw previously, ever since the 18th century, we have treated people like objects—as interchangeable soldiers in war, as chess pieces on the board to be moved around, as human resources, as human capital.

Our findings from surveying project professionals who are members of the Project Management Institute (PMI), Association for ement (PMI), and the Major Projects Association well as attendees at industry summits, conferences, asses, and webinars, have shown that project practitioners urning around people issues about six times more challenging an resolving technical problems. In other words, people capability and culture issues dwarf technical project issues by a factor of 6 to 1: people issues 84.8%, technical project issues 15.2%.

This should not come as a surprise. After all, project management is a human-centric activity. Project leaders are people. Clients are people (even if they are politicians, haha). Suppliers are people. Beneficiaries and end-users are people. Even organizations and companies consist of people. But herein lies the paradox: While people drive projects, the development of project professionals is heavily biased towards processes and systems. Relatively little attention is paid, for instance, to issues such as:

- Effectively building alignment on a project plan, or at least mitigating conflict among multiple stakeholders with diverging agendas;
- Managing suppliers to comply with the plan;
- Communicating with demanding clients;
- Turning around a toxic atmosphere and (or) mindset that has infected a failing project; or
- Overcoming cross-cultural differences leading to miscommunications, inefficiencies, and low productivity on a large, complex, geographically distributed project.

These are just a few examples, the proverbial tip of the iceberg. Human issues like these are the complex, tricky, annoying, irritating, and frustrating challenges project managers face daily. These are the issues where performance can be either recovered or lost—for good.

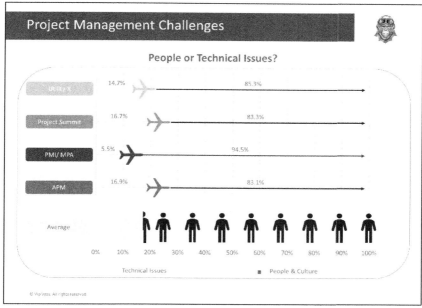

Figure 4.1: People vs. Technical Issues

The Human Element Drives Performance

From our five decades of aggregated experience, we have seen the good, the bad, and the horrible. We have worked with international clients to incubate the success of projects from the seed stage of an initial idea to navigating countless cliffs in the execution phase and the grind of turning around projects from the precipice of failure. Based on that experience, we assert the link between people and projects is robust and simple:

- When the project leadership is not aligned, the project will be fragmented.
- When the planning lacks clarity, delivery becomes chaotic.

- When the delivery team lacks ambition, the project lacks performance.
- When the project proponents lack trust, the project becomes contractual.
- When people are afraid to speak up, the project becomes risky.

We have repeatedly seen this dynamic changing dramatically when those who drive the project commit to altering the prevalent mindset and correlated behaviors. In these moments, the entire project shifts gears and moves to a new level of performance.

This is easy to say, especially in a book. But the ability to bring about this change in a live project environment requires vital capabilities not covered in standard project management textbooks or training. We call these capabilities "the art and science of megaproject leadership."

The good news is that they are capacities that you (and yes, we mean you, the person reading this right now or listening to it or staring at the e-reader screen) have the chance to build through this book.

At the risk of repeating ourselves: Our position is not to undermine the mechanical paradigm. It serves a purpose. But if we are to be effective at delivering million, billion, or trillion-dollar projects, we need to expand our field of vision and our dashboard of indicators. In the next chapter, we start this process by integrating the mechanical and people paradigms to capture and leverage the best practices from both worlds. To facilitate that integration, we offer a neat metaphor that, we hope, will stick because it is as simple as possible—but not simpler.

Radar Alert 5

THE TRAJECTORY OF YOUR ORGANIZATION,

OR PROJECT, IS SHAPED BY

THE TYPES OF THOUGHTS AND ACTIONS

IN 5 DISTINCT AREAS.

CHAPTER 5 > THE PROJECT FLIGHT PATH – A TOOL FOR SPOTTING THE GORILLA

All models are wrong.

Some are useful.

— George Box

Imagine boarding a plane from New York City to Rome, Italy. Also, picture the surprise on the passengers' faces when they discover upon disembarking that they've landed in Tokyo instead. Such deviations are highly unusual in the airline industry. Unfortunately, this is not the case for delivering projects where small and imperceptible jolts happen day-in and day-out. What do these tiny jolts look like? Here are some everyday examples:

- Participants in a project feasibility meeting *dramatically overestimating* the project's benefits.
- An engineer *keeping quiet* regarding a critical risk for fear of how his boss might react.
- Project leaders *ignoring resentments* between different functions or delivery organizations.

Each individual jolt may seem inconsequential.

Accumulated over time, the small can turn into the big, resulting in budget blowouts, tattered schedules, defective products, thwarted expectations, and failed ventures.

In our project analyses and interventions, we work with a simple metaphor: the Flight Path. Flights and projects share many characteristics. Both have a starting point A and a destination B, and both have a flight path to navigate from A to B.

The Project Flight Path views projects from two interrelated dimensions: project processes and people dynamics (see Fig. 5.1 below). The panel on the left ("Objective Dimension") maps out the key phases and procedures of the project; the panel on the right shows the subjective "below the surface" drivers of the human dynamics acting on the project. The depicted phases are consistent with the Project Management Institute (PMI) and the Association for Project Management (APM) frameworks. The content of the left panel can be replaced by other methodologies such as Agile or one that your organization uses. In our experience, this doesn't change the hidden patterns. The Project Flight Path allows for a robust process enabling projects to pull together the main phases and dimensions onto a single canvas.

Anatomy of the Flight Path

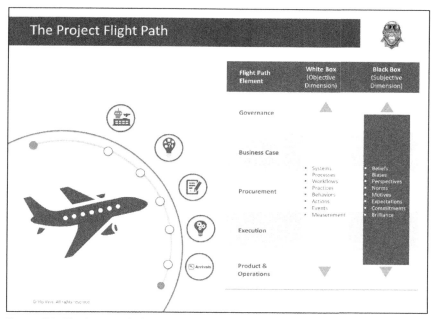

Figure 5.1: The Project Flight Path

Benefits of the Flight Path Model

The Project Flight Path tool is designed to get below the surface of project performance. The associated processes provide value-adds at every stage from A to B, whether the project is in the concept phase, the execution phase, or complete.

Project in Concept Phase (Design)	Project in Execution / Completed (Postmortem Debrief)
❏ Develop a shared visual picture of the project journey with critical stakeholders (investors, sponsors, delivery teams, key suppliers, etc.). ❏ Design and explore multiple project scenarios. ❏ Generate a higher-quality business case by improving the quality of reasoning and judgments on crucial project risks and challenges. ❏ Reveal potential blind spots and de-risk projects early on.	❏ Uncover the "project story" of both successful and failed projects. ❏ Enable users to see the big picture by simultaneously considering the impact of both visible and invisible dimensions on the project. ❏ Reveal deep project learning insights. ❏ Locate key project phases and areas where value was gained or lost during the project life cycle. ❏ Jettison what did not work/ standardize or scale up what worked. ❏ Examine, question, and validate the purpose and value of project practices and processes in meeting key objectives.

Table 5.1: Project Flight Path Value-Adds by Project Phase

We designed the Project Flight Path as a planning, diagnostic and intervention tool to help navigate projects by building successful flight paths. This includes being mindful of the "1-in-60 Rule" used in the aviation industry. The rule states that a one-degree off-track plane will be one mile off track for every 60 miles traveled. Preventing similar time- and resource-consuming deviations in megaprojects necessitates front-end quality conversations and high-value interactions. This is at the heart of spotting and pre-empting potential variations before the project swerves out of control.

Most projects don't fail for one reason alone. There isn't a single version of the truth regarding underperformance. Instead, there are multiple contributing factors. Consider that the shape of your project trajectory is the product of the interactions of five elements:

1. Governance, including competence and impact of stakeholders and political interests
2. Robustness of the business case
3. How well defined the product or deliverable is
4. Procurement philosophy and contracting models. and
5. Technical and social complexity of the tasks during delivery.

We hold the view, "Unless you can see it, you can't do anything about it."

Exposing the Patterns of Project Failures

Whether projects come under public scrutiny or not: what makes getting from A to B so tricky? We have asked this question in every project we turned around. We asked repeatedly and systematically. If you study any phenomenon long enough, you begin to see patterns in the data. Projects are no different: observe enough of them, and the template of failure becomes apparent.

On the surface, pick up the "lessons learned" from any project post-mortem, and you will see a repetition of the same clichés used to explain away poor performance. Typically, these issues are followed with a set of recommendations such as:

- "Improve Stakeholder Expectation Management."
- "Develop Estimation Process."
- "Tighten up Project Governance."
- "Assess Contracts for Optimal Application."

- "Review Project Controls (Schedule and Cost)."
- "Evaluate Supply Chain Performance."
- "Highlight and Address Project Capability Gaps."
- "Assess Safety Methods."

In the short term, such lists satisfy our psychological need for The Answer, but often these lessons evaporate from corporate memory. Think NASA: The famed US space agency produced historic space missions but failed to invest in its knowledge base.

When key project managers retired or resigned, much institutional memory, best practices, and lessons learned disappeared and were lost to the organization forever. The result: several failed missions, billions in sunk costs, and sometimes the terrible loss of human life.

Other organizations may face less dramatic consequences of such institutional memory loss, but the issue is the same: They repeatedly repeat the same mistakes.

Luckily, time and technology are on our side. Rapid advances in neuroscience, Artificial Intelligence, and Big Data are powerful "microscopes" that allow us to get massive data, examine projects at a level of detail previously unavailable, and reveal what was once the invisible part of the proverbial iceberg.

The Project Flight Path enables individuals to think deeper and visually isolate factors that collectively can create a toxic mix that poisons the project. We will follow the Project Flight Path by visibly dissecting the flight patterns of several well-known megaprojects whose budgets and/or schedules blew out of proportion and whose value evaporated. But first, we must delve into the hidden part of the iceberg and explore the messy human dynamics of megaprojects.

Radar Alert 6

EACH PROJECT AREA HAS TWO DIMENSIONS:

ONE TANGIBLE AND OBSERVABLE.

THE OTHER CONCEPTUAL AND INVISIBLE.

CHAPTER 6 > FLYING BLIND: WHITE VS. BLACK BOX

The greatest need for leadership is in the dark...

It is when the system is changing so rapidly...

that old prescriptions and old wisdoms

can only lead to catastrophe

and leadership is necessary

to call people to the very strangeness

of the new world being born.

— Kenneth Boulding

We hate to break it to you, but: The crash of flight ET302 we recounted in the Preface (if you didn't read it, go back there now), with its ensuing chain of fiascos at Boeing, was not an outlier. Far from it, the story of the flight reveals a systematic and repetitious pattern that runs in many projects. To illustrate, an example: Think about a recent project meeting with the team leader or a town hall with the CEO. The leader came in and perhaps showed a PowerPoint deck with a mission statement, or a senior manager came in and announced a key strategy.

We call this the "White Box " in keeping with our Flight Path model. This chapter will show you why.

The White Box

In the case of projects, the White Box consists of all the tangible, objective, and visible aspects of the projects, such as:

- The performance dashboards
- The business case
- The minutes of project board meetings
- The risk matrices
- The project schedule (GANTT chart)
- The Work Breakdown Structures (WBS)
- The engineering documents
- The tendering documents
- The construction drawings
- The safety processes
- The work method statements
- The user testing
- The quality checking and testing
- The defect reports
- The commissioning packs
- The ownership transfer documentation; and
- The reams and reams of other paperwork

You get the picture. What do you think: Do these items determine people's actions and ultimately results? Do they shape whether the project outcome is a success or a disaster?

This is not a rhetorical question.

It's the Black Box, Stupid

Then there is the background. Consistent with our Fight Path model, we call it the Black Box. It's what people in the meeting think but are not saying, "Oh boy, here we go again." "The boss never gets his hands dirty. He has no idea of the conditions on the ground." "How do I get out of this one? How can I cover my a__?"

It's what people whisper to each other at the water cooler, at the coffee machine or copier, after the meeting, or after the boss has gone: "They never consulted with us. We would have told them this whole thing would fall flat on its nose. Now they can eat it themselves." And often, it's what people say after work, in the bar when they have had one too many beers or vodkas or martinis, so we can't (really) print that here.

It's the Past still running and ruling in people's minds. "We tried the same thing last year, and it failed miserably. This will never work."

It's ultimately what the German philosopher Martin Heidegger called "what is un-thought in the thinker's thought": the phenomenon of obliviousness. Thoughts that are so far back in the recesses of their minds that they barely notice they are thinking those thoughts. Instead of saying, "People think those thoughts," we could almost say, "Those thoughts are thinking them." For example, at a major aerospace company we consulted with (no, not Boeing), we found, in a workshop after interviewing some 150 managers at all levels, that the Black Box was a conversation that went something like this: "Our best days are behind us."

Now that phrase sounds innocuous at first. But come to think of it, it's pernicious. Imagine coming to the office or project site every day, and the background hanging over you is, "Our best days are behind us." How is your work? What new frontiers do you envision? How do you respond to opportunities that emerge? Are you proactive or reactive? Even defensive? You know the answer.

It's not that people actively thought or said this sentence every day. It ran invisibly in the background. And that's precisely why the Black Box had such power over people's thoughts, actions, and ultimately the results. Now, just between you and us: Are your day-to-day behavior, your actions, what you see as possible, and your daily activities an expression of the mission statement the leader showed at the strategy meeting, or rather a reflection of the Black Box? Again, you know the answer.

Black Box vs. White Box

The nature of the White Box is very different from that of the Black Box. The foreground is visible and solid, whereas the background is invisible yet pervasive. The background is not something you put in place or choose. It's already there. It has always been there, ever since the project's inception and perhaps before. It's like the proverbial wallpaper: People don't see it anymore, even though it's right in their faces. And that invisibility in plain sight is precisely why the Black Box has the power to shape the project's culture, people's mindset, and what people see as feasible, possible, or impossible. It shapes people's attitudes and willingness to take responsibility or make commitments. It shapes their behaviors, whether they come to meetings on time, and whether and how they speak up in meetings. It shapes their actions. And hence it shapes their results.

If you prefer, another way to view project performance is as a projection of the Black Box, much like how a TV set works—the transistors and gadgets on the motherboard drive what you see on the screen. The screen is the White Box.

All organizations and projects have a White Box and a Black Box. And now comes the project killer: Traditional approaches to project management focus almost exclusively on the White Box and ignore the Black Box until it's too late. Or ever.

In the case of Flight ET302 Lion Air crashes, the dynamics taking place in the Background were at the heart of those fateful events. With ET302, the collective condition created in the background was instrumental in the final, terrible outcome. And here is the rub: Boeing's senior executives did not wake up every morning with the sole intention of creating a toxic work environment. That wouldn't make sense. After all, they had to rely on others for performance improvements and demonstrate quarter-on-quarter earnings growth to their shareholders.

In practice, the Black Box at Boeing developed over time, just like in every project or organization. It is also likely that Boeing executives were partially blind to the overall organizational condition they were fostering. It was a condition that overlooked shortcuts, tolerated cover-ups, put pressure on speed above safety, strangled dissent, blinkered individuals into cost-cutting at all costs, and promoted deception and deliberate manipulation. This, coupled with the downplaying of critical risks, led to the loss of 167 lives on that fateful flight.

Elements	Explanation
1. Project Governance	The system by which the project is controlled and handed over to operations. Governance includes how decisions are made, people are held accountable, and project performance monitoring.
2. Business Case	A business case is typically a well-structured document explaining the rationale for initiating a project, program, or portfolio. The content includes evaluating project benefits, an analysis of costs and risks, alternative options, and the preferred solution.
3. Final Deliverables	These are the outputs that clients expect once the project is complete.
4. Procurement Strategy	The procurement function's methods and procedures to effectively provide the project with the raw materials, equipment, industrial consumables, and services needed to complete the project's scope of work effectively. Procurement decisions also cover the delivery method used, e.g., Turn-Key, Design-Build, Design-Build-Operate, etc., and types of contract payment (fixed price, cost-plus, or other models) to incentivize performance.
5. Execution Capability	The critical skills and ability to execute the project's scope of work and deliver on the outputs expected by the client.

Table 6.1: The Project White Box ("Objective Dimension")

With reference to the Project Flight Path Model (Figure 5.1). whereas the left panel, the "objective dimension" is mechanistic and linear; the right panel is anything but. In the project's Black Box, we tap into the subconscious and unconscious root sources shaping human behaviors. This means delving into scientific disciplines such as behavioral economics, cognition, perception, biases, and heuristics—factors that determine the individual and group behaviors observed every day on projects. Unlike the panel on the left, the right panel is intangible, invisible, and lives "out of sight." We explore this invisibility, this remote part of the proverbial iceberg, in more detail.

How can you reveal what's in the Black Box? It's not easy. "Usually, people don't say anything," said Luc Gerardin, the transformation leader. "They walk instead. So, you have to listen; you have to show empathy. There are the official conversations, and then there are the unofficial conversations." The conversations that determine the outcomes, the successes, and failures of projects, are not the official ones you hear in the press releases or the town halls.

Gerardin, who worked with IBM on several megaprojects, compared the mindset prevailing in high-tech projects of Silicon Valley with the mentality of other sectors, for example, logistics. "It's all about how you run the company: agile, not rigid. Today's most valued companies are the most agile: Alphabet, Amazon, Apple, Meta." But shipping is a man's world. "They usually don't have a win-win mindset, receiving, flexibility, or building a shared vision. Instead, they have a misalignment of agendas."

How do you go about revealing the Black Box when the "corporate immune system" guards vigorously against change? "The more transparent you are, the more effective you will be. You put on the table what you want, your fears, your doubts." And you must pick up on the small signs. How are people rationalizing their behaviors?

What is their underlying belief system? For example, one fundamental worldview in megaprojects can be "You or Me."

"People are usually in a zero-sum game mindset: If you win, I lose," Gerardin said. "But if people focus only on their gain, the sum of payoffs is lower than with win-win. If you don't work against each other, everybody wins for everyone's benefit. It's all a question of mindset: in this case, a partnership mindset instead of a mindset of competition, 'I win you lose,' winners and losers." Unearthing such fundamental dynamics is hard work. "But if you succeed, the benefits are huge."

The Future Sits in the Black Box

"The fourteen people involved were very significant--bright, able, dedicated people, all of whom had the greatest affection for the U.S.... If six of them had been President of the U.S., I think that the world might have been blown up." [20]

Robert F. Kennedy was, of course, looking back at one of the defining moments in U.S. history: the 1962 Cuban Missile Crisis, when U.S. intelligence agencies discovered the Soviet Union had stationed ballistic missiles in Cuba in the proverbial backyard of the U.S.

When you look back in history, it seems as if the course of action that a historical leader, in this case, John F. Kennedy, chose was the only possible one. But when you are present at the creation, nothing is inevitable. There are many options, people are vehemently arguing for their pet option and disparaging alternatives, and tiny details like who happens to be in the room can decide the nation's fate.

To understand why the United States, after two frantic weeks of hand-wringing and fierce debates, finally moved to blockade Cuba on 24 October 1962, we need to look at more than just the minutes of the meetings or the foreground. As Harvard professor Graham Allison wrote in his brilliant analysis of the crisis: "The American government's choice to mount a blockade cannot be understood apart from the context in which the necessity for choice arose."[21]

There was the timing: Had an American U-2 flown over the western end of Cuba two or three weeks earlier, it would have discovered the missiles. There was the fact that Cuba was the Kennedy administration's "political Achilles' heel" after the previous year's failed Bay of Pigs invasion. There was a public opinion: The failed invasion had made Kennedy look indecisive. There was also the upcoming Congressional election and the declaration by opposition Republicans to make Cuba "the dominant issue of the 1962 campaign."[22] When the President and his team of advisors weighed possible pathways out of the crisis, each factor was integral to the context that influenced their choices.

A possible outbreak of World War III is a dramatic example, but the same thing happens in megaprojects. Whatever lies hidden in the background is the ultimate root cause of outcomes in performance. Let's see how this Black Box might play out in a project. Picture Edmund, a Chinese contractor whose company is responsible for supplying the foundations for an upmarket shopping mall. The mall is to house hundreds of luxury stores, premium restaurants, funky coffee shops, a large food court, a state-of-the-art 4DX movie theater, and a bowling alley. The shopping mall even has an indoor aquarium with a mini-submarine that kids can climb into.

For those unfamiliar with construction, foundations can be thought of as the "legs" of a building, bearing its weight. And given the enormous mass of the shopping mall, the legs needed to be strengthened using concrete, a common technique used in construction.

Today Edmund has a meeting with the client to give a progress report. Just before the meeting, he got a call in the car. The news couldn't be worse: One of his main sub-contractors, responsible for pouring the concrete for the basement, has gone bust. Multiple thoughts tumble around in his head: How can he find another sub-contractor immediately? Will they exploit his predicament and charge a higher price simply because he has his back against the wall? And even if he does manage to get a sub-contractor, they probably won't know the ropes. How can he onboard them quickly without jeopardizing the project delivery schedule? And right now, when he walks into that conference room, how is he going to break the news to the client—who frankly has not been the easiest to work with, to put it mildly, as it is.

With fear and trepidation, he walks into the client's office. The project is already three months behind schedule, and the project atmosphere is touchy and tense. Centre stage is Brad, an aggressive Australian Project Manager, flying off the handle. "What the f*ck is this?" he says as he thrusts a pile of engineering drawings back at the Design Consultant. "It's a pile of sh*t. Call your boss and get him in here right now."

As the expletives bounce off the walls, everyone (other than Brad) keeps silent, waiting for the storm to pass, hoping they won't get struck by lightning. The discomfort is palpable. The room is defensive. People are struggling to gather their thoughts. Silence and self-preservation seem to be the ideal strategy for everyone. And Edmund is no different.

Ok, so Brad may be an extreme example, but from our experience, it's not unusual for project leaders to "snap" when the pressure cooker starts to simmer.

The Linchpin Factor: Leadership

Leadership plays a critical role in shaping the Black Box for megaprojects and corporate change initiatives. Why? Simply

because we all have a systematic
We expect them to set the tone
characteristic we have inherit
humans, we look for role mod
tantrum creates a backgroun
protection: a perfect cockt
news up the chain of command, lea

Even if you are not like Brad, as a
behaviors, conversational tone, and habits have mu
than you might imagine. The truth is, you are on "sho
everyone is looking at cues to guide their behavior. In some cases,
even simple mimicry, such as raising the occasional eyebrow or
frowning, or speaking loudly, is all it takes to have the brains of your
employees and contractors primed for survival. Their limbic
systems get activated, a "Threat State" is switched on in their minds,
and survival measures kick in.

When people are in survival mode, you can expect to see the
following: They keep quiet. They don't speak up on critical issues.
And they worry about what others will think of them (a.k.a. social
survival). In some cases, they secretly contact a headhunter to look
for greener pastures: for job opportunities that are less angst-
ridden, more compelling, a respite from the toxic project
environment they are currently enduring (and that you might very
well have had a hand in creating).

So, at the risk of repeating ourselves, the quality and
performance of the White Box—whether a project meets success or
failure—depends on what happens in the Black Box. But how do we
open the Black Box and peek inside?

"I was part of the merger between CMA and NOL APL in Asia,"
recounted the megaproject consultant Luc Gerardin. The $2.4 billion
corporate marriage between two shipping companies was quite a
culture clash. "One culture was very corporate; the other was flat

urial. Usually, the startup suffers from the merger. e buys startups as an investment to change its own

ning the Black Box: The vers and Dials of Projects

Stop and pause for a moment to review the list of beliefs below. Which of these resonate with your experience in your megaproject (or portfolio of projects)? Where does your project sit on the spectrum? Next, assess the impact these beliefs have on performance areas such as quality of the business case, governance, procurement, delivery effectiveness, task ownership, amount of rework, costs, schedule, and stakeholder satisfaction.

Performance Success Factor	Constrains	Expands
	<------ Spectrum of the Black Box ------>	
Leadership	"She's the boss..." "My hands are tied..." "This issue is above my pay grade..."	"I connect with the right people and make sure the project tackles this."
Motivation	"This project is just a job.." "I'm just a cog in the wheel...."	"We're re-designing the economy and future of X."
Trust Barriers	"The client/contractor/ government is out to shaft us..."	"We're both in a difficult situation. Let's see what we can work out."
Fear of Speaking Up	"We can't possibly say that or raise that issue..." "Don't rock the boat...."	"Let's create an amnesty to get on the table what's going on."
Accountability	"It's not my job; someone else will deal with it..." "Let's hope for the best...."	"The buck stops with us."
Culture	Dead silence in the room...	People contribute and generate practical insights.

Table 6.2: The Project Black Box, Sample Background Conversations

Have you ever sat in a meeting where the real issues weren't discussed? Where an undercurrent of fear prevented people from addressing the proverbial elephant in the room? Later in the book, we will show you how this plays out in real megaprojects.

From Black Boxes to Black Swans

We both experienced our share of Black Swans.

It was a beautifully clear day in September 2001, an Indian summer morning when the sky was deep blue. Thomas was sitting on the Brooklyn Promenade--alone except for a few runners and dog walkers--and reading Michel Houellebecq's *Les Particules Élémentaires* (this is not meant to be an endorsement of that book, far from it) when he looked up at 8:46 a.m. and saw something he had never seen before: A plane hit the World Trade Center. Smoke and millions of tiny metallic glitters were in the air; a light wind swept them toward him. The glitters were countless papers and documents flying across the East River. One of them was a page from a civil law book, blackened on all four sides. Another was a FedEx envelope with a contract someone had just signed a few minutes earlier.

About a half-hour later, another plane flew in from Staten Island, right over the Statue of Liberty. It flew low and accelerated head-on toward those now gathered at the Promenade. It banked like a fighter plane, its dark underbelly visible—a terrifying sight that you usually see only in war zones or movies. Suddenly the aircraft ducked behind a skyscraper and disappeared into the South Tower a moment later. About a dozen people were watching, speechless and transfixed by this time. Thomas called as many people as he could on his mobile but got through only to his parents' answering machine in Sydney before his phone went dead. Then he saw one tower collapse, then the other. His knees gave in; he staggered to a bench, sat down, and wept. It was hard to breathe.

Like so many others, the only productive thing Thomas could think of was donating blood. It seemed like a drop in the bucket. 9/11 was an event nobody (to our knowledge) had predicted. For days, the United States government would be caught off guard. Nobody knew where the president was; the vice president was at a "secure undisclosed location." The skies above were empty. The New York Stock Exchange and the Nasdaq were closed through 17 September, the most extended shutdown since the Great Depression. The nation was at a standstill.

Fast forward 18 years: Vip was in Hong Kong in early December 2019 when news broke of a deadly bug having jumped from bats to humans in a wet market in Wuhan, China. He didn't think much of it, other than the thought that "Wuhan is a long way from Hong Kong." He was preparing for the annual Christmas break, reconfirming flights and hotels for the family's next overseas trip to Patagonia. Two weeks later, the family flew to Buenos Aires—only to sit in the hotel lobby, watching CNN, and the seeds of fear and panic spread across the globe. In late January 2020, the Hong Kong Vip's family came back to was entirely different from the one they had left five weeks earlier. Eerily quiet streets now replaced a city known for its electric buzz. Sporadic panic buying laid supermarket shelves bare of the essentials. But worse of all was that the soul of Hong Kong had been ripped out. Warmth and friendship were replaced by suspicion, distance, and alarm.

As his company reconnected with its international clients, the message was the same: The Covid-19 pandemic had impacted every aspect of their businesses' Flight Paths. Since shutdowns were in place, most projects had to contend with inaccessible regulators, and critical approvals had gone into "suspended animation." Pivotal go/no-go decisions were being delayed. Projects also faced other issues such as supply chain disruptions, logistic backlogs, falling production rates, site stoppages, anxious workers, frightened families, remote working, and other government restrictions. Worried company leaders had to get on top of entirely new challenges that most (if not all) were unprepared for.[23] Nobody could have predicted we would be hit by a global economic disruption that would last for years—and is still ongoing as we write.

Whether they take the form of pandemics or terror attacks like 9/11, stock market crashes like the 1987 Black Monday, or Singularity (the tipping point where technology growth becomes irreversible and gets out of control of humans, threatening humanity with extinction), Black Swans are highly unpredictable, rare events with the potential to cause catastrophic damage to the economy, and our existence.[24]

Unsurprisingly, this includes projects. While this book is not directly about Black Swans, their impact on Flight Paths can hardly be overstated. And how we respond, as you may have guessed, comes down to the Black Box.

Bottom Line: The White Box (the project's visible or technical foreground) is not what derails most projects. It is in the Black Box (the project's background) where all the processing, biases, heuristics, and secret talk happens. It is where the organization's beliefs, biases, and (sometimes untapped) brilliance live. The Black Box is where the wide range of hard and soft project data and information is received, processed, and actioned. The brains and minds are the resource centers of your project. In the background, we find a whole range of mental dynamics and barriers determining the performance of the foreground or White Box.

How the brains and minds of those participating in the change effort or project are configured and oriented makes all the difference to project success (or failure). And project leadership has a big say in how this is designed.

Here is a single, particularly potent question worth considering: As a project leader, are you creating the conditions that foster imagination, creativity, communication, collaboration, action, and performance? Or are you directing the project's most vital resources (the people) towards worry, fear, blame, excuses or self-protection?

To lead a project successfully requires effective leadership and management of both the Foreground and Background. Two reasons. First, take an organization that refuses to change. Who do you think has the most incentives for keeping the status quo? And who is (usually) most threatened by change or innovation?

You guessed it: top management.

Radar Alert 7

THE BLACK BOX, AT ITS CORE, HOLDS THE "3 BS":

BELIEFS, BIASES, AND BRILLIANCE.

BELIEFS SHAPE THE ACTIONS

THAT PRODUCE THE OUTCOMES.

EVEN WHEN THE BELIEFS ARE WRONG.

Chapter 7 > Mini-Case: The Black Box's Explosive Impact on Deepwater Horizon

On 20 April 2010, a series of explosions shook the BP Macondo oil well in the Gulf of Mexico to its core. And not only the well: The worst blowout gutted the nearly 400 feet long Deepwater Horizon stem to stern. Crew members were cut down by shrapnel, hurled across rooms, and buried under smoking wreckage. Some were swallowed by fireballs that raced through the oil rig's shattered interior. Dazed and battered survivors, half-naked and dripping in highly combustible gas, crawled inch by inch in pitch darkness, willing themselves to the lifeboat deck.

But it was no better there.

That same explosion had ignited a firestorm that enveloped the rig's derrick. Searing heat baked the lifeboat deck. Sure they were about to be cooked alive, crew members scrambled into enclosed lifeboats for shelter, only to find them like smoke-filled ovens. Men admired for their toughness wept. Several said their prayers and jumped into the oily seas 60 feet below.

An overwhelmed young crew member, Andrea Fleytas, finally screamed what so many were thinking:

"We're going to die!"[25]

The Deepwater Horizon drilling rig catastrophe was a night of unprecedented horror, with 11 crew members killed and 17 seriously injured. It is considered the worst ecological disaster the USA has ever experienced. Images of the flotilla of boats battling to contain the raging inferno were beamed worldwide, and the whole incident cost then-CEO Tony Hayward his job. The blowout altered and destroyed many lives forever, not to speak of the oil giant suffering heavy penalties (some $70 billion in US federal fines). There was further fallout for BP: massive lawsuit payouts to Gulf Coast businesses and residents, a severe blow to its reputation, and a crash of its shareholder value by half, so far permanently (a decade later, BP's share price is still half what it was pre-disaster). But by far, the worst was the loss of human lives.

- Jason Anderson, senior tool pusher.
- Dale Burkeen, crane operator.
- Donald Clark, assistant driller.
- Stephen Curtis, assistant driller.
- Gordon Jones, mud engineer.
- Wyatt Kemp, derrickman.
- Karl Kleppinger, roughneck.
- Blair Manuel, mud engineer.
- Dewey Revette, driller.
- Shane Roshto, roughneck.
- Adam Weise, roughneck.

We owe it to the men who died and to their families (not to mention the wildlife and environment that were destroyed irreversibly) to find out what happened and to reveal the causes, so it never happens again.

Postmortem: Inside the Blackbox

On the surface, it looks like a double disaster—first the blowout, then the destruction of the Horizon due to a series of technical failures. But now, looking back twelve years after the disaster, we can reveal the long-term problems hidden by the traumatic events at the time.

The Horizon had formidable and redundant defenses against even the worst blowout. It was equipped to safely divert surging oil or gas away from the rig. It sported devices to quickly seal off a well blowout or break free from it. It had systems to prevent gas from exploding. It had sophisticated alarms that could quickly warn the crew at the slightest trace of gas. The crew itself routinely practiced responding to alarms, fires, and blowouts. It was blessed with experienced leaders who cared about safety. [26]

The Horizon's safety performance was so good that the month before, two BP executives and two officials of Transocean, the world's largest operator of offshore rigs and the Horizon's owner, flew out to the rig to praise the team's safety performance. And the Horizon's chief, Jimmy Harrell, handed one of the crew members a handsome silver watch in recognition of an excellent rig inspection.

So on paper, the Deepwater Horizon should have weathered this perfect storm. But it didn't. Why? What was the source of the disaster?

To start with, there was something off in the relationship between BP and the Transocean crew. It was hard to put the finger on it, but the VIPs did not just fly out to praise. BP managers—whose bonuses were heavily based on saving money and beating deadlines—kept asking in emails when the well would be finished. It was obvious that the guys in the drill shack felt the executives from BP were breathing down their necks. The writing was on the wall, Caleb Holloway, a 28-year-old floorhand whose nickname was Hollywood, recalled later. "You could just tell."

BP denied pressuring the Horizon crew to cut corners, but it must be said that its plans and deadlines for completing the well kept changing, often in ways that saved time but also increased risk. "It's a new deal every time we wake up," Jason Anderson, a 35-year-old senior toolpusher who died in the disaster, had complained to his father.

That fateful evening, the rig crew and the BP people had to do one final, crucial test before the Horizon could plug the Macondo well and move on. To ensure the well was not leaking, the crew would withdraw heavy mud from it and replace it with lighter seawater. Then they would shut in the well to see if pressure built up inside. If it did, that could mean hydrocarbons — oil and gas — were seeping into the well. In effect, they were daring the well to blow out. Designing the test was BP's job, yet the oil company's instructions, e-mailed to the rig that morning, were brief: 24 words, about the length of a Tweet.

It fell to the BP company men and the drilling crew to work out the details, but it did not go well. There was strong disagreement over the test results. A BP company man said he thought the test went fine. He was not alone. "The Transocean rig crew and BP well site leaders reached the incorrect view," the BP internal investigation team was to write, "that the test was successful and that well integrity had been established."[27] But in reality, pressure had built up, precisely what they did not want, and Wyman Wheeler, the team's toolpusher, was worried.

In seven years on the Horizon, Joseph Keith had never seen so much activity while sealing a well, making him uncomfortable. His job included monitoring gauges that detect blowouts. But all the jobs going on at once — transferring mud to a supply vessel, cleaning mud pits, repairing a pump — could throw off his instruments. Mr. Keith did not tell anyone that he was worried about his ability to monitor the well. "I guess I just didn't think of it at the time," he later testified.[28]

Was the crew distracted by other tasks, rushing the job, or simply complacent? We will never know. But investigators would later agree that the Macondo well had failed its crucial final test, with fatal consequences. "The question is why these experienced men out on that rig talked themselves into believing that this was a good test," said Sean Grimsley, a lawyer for a presidential commission investigating the disaster. "None of these men out on that rig want to die."[29]

But there were tell-tale signs before the fateful night that something was amiss. Take the blowout preventer, a 400-ton monster machine that was the crew's mightiest weapon. It gave the men several different methods to shut in the well, the most extreme being a robust set of hydraulic shears that could cut through a drill pipe and seal the well.

The industry has long depicted blowout preventers as "the ultimate fail-safe." The evidence shows a different picture: The blowout preventer may have been crippled by poor maintenance. Investigators found many problems —dead batteries, bad solenoid valves, leaking hydraulic lines — that were overlooked or ignored. Transocean had also never—not even once—performed an expensive 90-day maintenance inspection that the manufacturer said should be done every three to five years. Industry standards and federal regulations said the same thing. BP and a Transocean safety consultant had pointed out that the Horizon's blowout preventer, a decade old, was past due for inspection. But Transocean decided that its regular maintenance program was adequate for the time being.

To be sure, the drill crew had trained for blowouts. Floorhands like Caleb Holloway were the crucial first responders. A driller would call "Blowout!" and time the crew's response. A special valve would be mounted on the drill pipe to stop the imaginary blowout as quickly as possible. But now it was real, and Caleb Holloway realized there were no floorhands on the drilling floor to respond. Holloway cursed and sprinted for the stairs. Dan Barron, the crew's newest member, was right behind him. A waterfall of mud was spilling from the drilling floor to the main deck. Six meters from the blowout's full fury, it sounded like a jet engine, a shrill whining howl. Holloway was instantly soaked, his protective glasses coated with mud.

He reached for his radio and called the driller, Dewey Revette. "Dewey!" he shouted. "What do I do?" He didn't hear an answer. He pulled out his earplugs and tried again. "Dewey. Dewey. What do we need to do?"

Again, no response.

Doug Brown and his men were right next to the emergency shutdown system. All they had to do was to lift the plastic cover, hit the button, and all engines would shut down in seconds.

But they did not.

And to be fair, the step was neither obvious nor easy to take. There was a risk in overreacting. If they killed the engines, the Horizon would drift from its position over the well, possibly damaging the drilling equipment and forcing costly delays. Indeed, Doug Brown would testify later that he did not think he had the authority to hit the emergency shutdown. The practice was to "wait and listen" for instructions from the bridge, said William Stoner, one of the men with Brown. But other than the two brief calls, each only seconds long, there were no communications or coordination among the bridge, the drill shack, and the engine control room. The men in that room did nothing.

According to Transocean, it was a full six minutes after the mud had been gushing onto the drilling floor when the crew finally hit the general master alarm. Government officials and BP's internal investigation revealed it was even nine minutes.

There was still one final way to keep the Horizon from sinking. One of the supervisors responsible for the blowout preventer, Chris Pleasant, saw it first. With the main deck on fire, he ran for the bridge with one thought: They needed to disconnect the rig from the blowout preventer and the well itself. That would cut off the fire's primary fuel source and give the Horizon a fighting chance. He just needed to activate the emergency disconnect system (EDS). Like a fighter pilot hitting eject, it would signal the blowout preventer to release the Horizon. It would also signal it to seal the well, perhaps stopping the oil flow into the Gulf of Mexico.

"I'm hitting EDS," he told the captain.

Witnesses differ about what happened next. But they agree on an essential point: Even with the Horizon burning, powerless, and gutted by explosions, there was still resistance to the most decisive measure that might save the rig. According to Mr. Pleasant, the captain told him, "No, calm down; we're not hitting EDS."[30]

"Multi-Decade Organizational Malfunction and Short-Sightedness"

In its conclusion, the Deepwater Horizon Study Group (DHSG) highlighted organizational attitude and decision making as crucial contributing factors to the disaster: "...these failures (to contain, control, mitigate, plan, and clean-up) appear to be deeply rooted in a multi-decade history of organizational malfunction and short-sightedness."

The DHSG's report could scarcely have been blunter: "management's perspective failed to recognize and accept its own fallibilities despite a record of recent accidents in the U.S. and a series of promises to change BP's safety culture."[31]

For example, communication fell apart. Crew members hesitated at crucial moments and shied away from necessary actions. Warning signs were missed. Crew members in critical areas failed to coordinate a response.

The DHSG found that "BP's system 'forgot to be afraid." Its final report did not mince words: "poor decision-making played a key role in accident causation."[32] And why was decision-making poor? The inconvenient truth: Just like with the Boeing disasters earlier in this book, the root cause was cutting corners for the sake of profits over safety:

> Analysis of the available evidence indicates that when given the opportunity to save time and money—and make money—tradeoffs were made for the certain thing—production—because there were perceived to be no downsides associated with the uncertain thing— failure caused by the lack of sufficient protection. Thus, as a result of a cascade of deeply flawed failure and signal analysis, decision-making, communication, and organizational-managerial processes, safety was compromised to the point that the blowout occurred with catastrophic effects.

The report dug even deeper and found a pattern of cultural blind spots that operated at BP:

> At the time of the Macondo blowout, BP's corporate culture remained one that was embedded in risk-taking and cost-cutting—it was like that in 2005 (Texas City), in 2006 (Alaska North Slope Spill), and in 2010 ("The Spill"). Perhaps there is no clear-cut 'evidence' that someone in BP or in the other organizations in the Macondo well project made a conscious decision to put costs before safety; nevertheless, that misses the point. It is the underlying "unconscious mind" that governs the actions of an organization and its personnel. Cultural influences that permeate an organization and an industry and manifest in actions that can either promote and nurture a high-reliability organization with high-reliability systems, or actions reflective of complacency, excessive risk-taking, and a loss of situational awareness.

The DHSG's findings were right on the money. The problem with its report is that its seven recommendations focused exclusively on technical solutions and changes in governance and oversight. These factors doubtless played a role; they are necessary but not sufficient, not by a long shot. While the report correctly highlighted the crash factors of what the DHSG called "unconscious mind" and "deeply flawed failure and signal analysis, decision-making, communication," its prescriptions themselves fell victim to the same blind spots—they were blind to these factors. This made sure that the next disaster would loom inevitably on the horizon. "Those that fail to learn from history," Winston Churchill famously said, "are doomed to repeat it."

The Tip of the Iceberg

Alas, accidents of this scale are not new in the extractives industry. In many ways, the Macondo disaster closely replicates other catastrophes in offshore oil and gas. Texas City, Exxon Valdez, Piper Alpha, the Montara well off the coast of Australia in the Timor Sea that occurred a mere eight months before the Macondo well blowout, and others still linger in our memories.

Oil and gas is only one area, albeit a spectacular one, in which megaprojects crash. As we have seen, they crash all over the place, but we usually don't hear about them. Take the giant area of US Homeland Security. In Afghanistan for example, successive U.S. administrations, both Republican and Democrat, have spent $5 billion in taxpayers' money *a month* over at least a decade. The neighborhood of Wazir Akbar Khan (named after the general who commanded the Afghan army's rout of the British in 1842) is a ghost town of luxurious, giant but dilapidated mansions nobody lives in anymore because they took their families to Dubai or the United States, or some other place that is safer for them and their money. The streets are strewn with potholes. The sewage runs down the sidewalks. The trash has not been picked up. "It is difficult to conclude," said the U.S. Special Inspector General in 2011, "that the project was worth the investment."[33]

Is it possible that all these failures of megaprojects stem from one source, one root cause? In the next chapter, we will show you how to peek into the Black Box.

Radar Alert 8

ZOOMING INTO THE BLACK BOX REVEALS

THE MACHINERY BEHIND OUR BEHAVIORS.

Chapter 8 > Invisible Project Killers: Neuroscience and the Black Box

We don't see things as they are.

We see them as we are.

— Anaïs Nin

The Project Management Institute (PMI) defines a project as a "temporary endeavor undertaken to create a unique product, service or result." The <u>keyword</u> here is "temporary": Temporary distinguishes a project from an ongoing business with continuous operations and repeated activities. A major or megaproject can be located somewhere along a continuum between the hierarchical firm at one extreme of the spectrum and the decentralization of the market at the other.[34] It's a bit like making a movie, where the producers pull together multiple stakeholders for a limited time to make the picture and bring it to the box office.

A Stressful Definition

Temporary means "lasting for only a limited period of time." While the definition is helpful, what is less obvious is the impact the word "temporary" has on your brain. Notice the different reactions you have listening to the following sentences:

- "This is a temporary job for three months" versus "This is a permanent role."
- I am not sure if we're compatible; let's have a temporary relationship" versus "I think you are awesome. Would you be interested in something more solid?"
- "You have 5 hours to turn around this temporary report" versus "This report is due the week after next."

Within the context of projects, "temporary" means a lack of continuity, of sustainability, and of reliability. It can bring urgency and time pressure to complete tasks within narrow time frames. It can also generate anxiety since those delivering the tasks contractually often face negative consequences if things don't go according to plan. These can take the form of penalties and fees ("Liquidated Damages") and even the prospect of losing one's job if tasks are not completed to the required specifications.

How the Brain Deals with Urgency

Now how does urgency play into that? The impact of speed and urgency on our brains has been explored by Alberto Megías and his colleagues using a series of experiments focused on the task of driving a car under a range of risky (stressed) and non-risky (stress-free) conditions[35]. An example of the former is braking hard when a ball suddenly appears between two stationary cars on the road. Often, the ball is followed by a child chasing it in this scenario.

The researchers found that urgent behaviors were strongly correlated with activity in the anterior cingulate insula, an area of the brain associated with processing emotions, and regions of the frontal lobe, activated by the identification of threats.

The Activation Map of the brain under the conditions during the experiment is shown in the image below. Notice the dramatic visual difference between the two states: urgent versus evaluative.

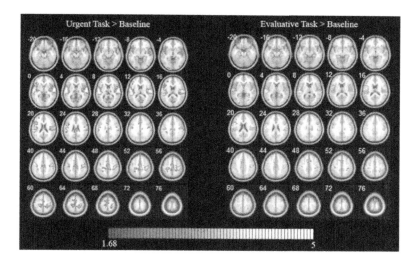

Figure 8.1: Activation map for the "urgent>baseline" and the "evaluative task baseline" contrasts in orthogonal projection

The Filter of Fear

If we can believe the statistics, it is well-known that men think about sex every 7 seconds. That sounds amazing until you look at how often the brain assesses the environment for threats. How often, do you think?

Every 1.6 seconds.

Every 1.6 seconds, the brain checks: "Am I still safe? Am I still safe? Am I still safe? Any threats coming my way?" It's our survival mechanism. To be sure, that mechanism has been precious. Without it, we probably would not have survived as a species. Some 100,000 years ago, when you crawled out of a cave and suddenly you felt a shadow above you, you had to decide in a split second: Is this a condor about to swoop down and have me for lunch? Or is it just a lovely cumulus cloud overhead? In the anarchic world at the time, a world without institutions or rules, when you met a stranger, you had only seconds to decide whether that stranger was a friend or foe. Your survival depended literally on being paranoid.

Remember Andy Grove, the former CEO of Intel? You may or may not remember the title of his memoir. It was called *Only the Paranoid Survive.* As a Holocaust survivor, Grove's attitude was understandable.

Our brain deals with the signals coming in through the five senses of seeing, hearing, smelling, tasting, and touching by channeling the incoming data traffic through the amygdala, our internal danger detector. Part of our limbic system, the amygdala constantly scans for dangers and threats. It is hard-wired to spot even the slightest threat to our survival, pre-readying the body for three reflexes: flight, fight, or freeze.

There is a tug-of-war between the amygdala and the Rostral Anterior Cingulate, part of the brain's frontal cortex. And usually, the amygdala wins. An example of this is when we watch a politician on TV, one we have never seen before, and decide within a second, without knowing anything about that politician's ideology or program, whether or not we will vote for them.

The fear reflex is deeply embedded in our amygdala (also called the reptilian part of our brain). Besides priming our brain and body for impulsive danger avoidance, the amygdala is also an emotional center responsible for generating other emotions such as anger, stress, disgust, or upset. This is sometimes described as the "Amygdala Hijack." And after the hijack has passed, other emotions like embarrassment, regret, or resentment might flood the brain. The amygdala is the seat of the deep unease you experience when you are in a meeting and your boss looks over and gives you the "dirty look."

Your racing heart, muscle tensing, and sweat production are symptoms of the amygdala activating the sympathetic nervous system, telling the adrenal glands to inject more adrenaline, and readying your body to spring into action.

A critical characteristic of the amygdala is that it cannot distinguish between real and perceived or probabilistic threats. Probabilistic risks are quite different from real ones, with a definite start and stop in time. But your amygdala doesn't know that. It responds in the same way. This is why you may find yourself closing your eyes or hiding behind the sofa when a terrifying scene comes on screen. While you know the woman chasing the man with a meat cleaver are both actors, that logic makes little difference when you watch the woman closing in on the victim.

When Thomas' mother grew up during World War II in Basel and heard the bombs fall just beyond the German border every night, it made no difference to her amygdala whether the Gestapo was about to march into her bedroom or whether this was just fear or fantasy. And it makes no difference to your amygdala if your boss will fire you or if you're just scared that he might.

The world of projects is filled with probabilistic risks and threats. Hence a key focus in project management is risk identification and mitigation. It is not uncommon for projects with risk registers to list hundreds, if not thousands, of identified probable risks.

While listing and evaluating risks is a prudent act, another feature of probabilistic fears is worth highlighting—probabilistic worries never go away. As we have highlighted, "Something bad could always happen," or "You'll see. You may get caught at the next turn." In the context of projects, concerns associated with perceived or probabilistic risks mean that our brain is continuously on edge.

For example, notice the continuous anxiety levels associated with "The system network could go down" compared to "We have identified a bug in the software that we need to correct by Friday."

No wonder then that fear is rampant in complex projects riddled with uncertainty and multiple variables. After all, the tsunami might hit at any moment.

Besides dealing with urgency and compressed timeframes for task completion, and the filter of fear, our brains face multiple other obstacles in major projects. One of them is the tidal wave of incoming signals.

Information Wipe-Out

The problem of data and information overload has been studied extensively by information scientists. Many have commented on how advances in digital technology over the last 50 years have rapidly outpaced the brain's evolutionary development to process that information over the same period. For example, did you know that in 2020, YouTube users uploaded 500 minutes of video *per minute*, and WhatsApp users now send 41 million messages *per minute* (and counting.) By 2015, we had already created a world with 300 exabytes of human-made information. That's the number 3 followed by 20 zeros, an unimaginably large number. For illustration, your standard 90-minute feature film is about 1 Gigabyte. If we take that number, 300 exabytes equals 322,122,547,200 Gigabytes, to be exact; more than 322 billion movies. If you watched one movie every night, as some of us do, it would take you about 882.5 million years to view all these movies. Talk about binge-watching.

Studies by the psychologist Mihaly Csikzentmihalyi and independently by Bell Labs engineer Robert Lucky calculated the "bandwidth" of our conscious attention to be 120 bits per second;[36]

a tiny fraction of the information coming our way since the human body sends some 11 million bits to the brain every second.[37] (A newer estimate by Fermin Moscoso del Prado Martin estimates the capacity our brain can process during lexical decision tasks at only 60 bits per second, which makes the problem even worse. Either way, you get the picture.)

From a project perspective, every day, news about stakeholder dissatisfaction, political pressures from the top, investor concerns, regulatory scrutiny, design flaws, logistical holdups, production challenges or delays, commercial in-fighting, employee dissatisfaction, and cross-cultural stand-offs, to name just a few, have to be channeled through this very narrow bandwidth by the leadership teams. Or, to make it more personal, by you. (Do you have a headache yet?)

All of the sensing, detecting, processing, and acting is done by our neurons. And given the daily surge of information that rushes to our shores, the brain and nervous system expends a lot of energy to wade through, process, organize, and make sense of the volume, variety, and volatility of inputs it receives. No wonder so many of us are tired, no: exhausted at the end of the day. As our glucose levels—the primary fuel for the brain—run down, so do we.

Protecting Self-Image from the Ultimate Threat

In the business context, how often have you heard people say, "I thought I was gonna die," or "They burned me at the stake," or "She cut my balls off and then handed me to the firing squad" after they came out of a tough meeting?

Phrases like these mean that the speakers' identities (or egos) survived the experience. And indeed, for an individual, a direct threat to one's identity is about as dire as it can get! It is also the source of many nasty, counterproductive behaviors you see when projects become distressed and the room is filled with anxious and unsettled stakeholders.

Repeating Mistakes Over and Over—Why?

Our experience working on many megaprojects and change initiatives is that organizations tend to repeat the same mistakes time and again. Sometimes this can be "chalked off" as institutional memory loss since delivery teams disband and disperse onto other projects or move to other organizations. And, their knowledge goes with them or vanishes altogether.

This sounds like a valid explanation. Tacit knowledge acquired during delivery is never systematically captured and is readily lost. But this is not the whole picture. The logic doesn't hold. If it did, the very same managers who move on to new projects would not repeat the same patterns of judgments and errors. But they do.

In a recent study published by *The Journal of Psychiatry and Neuroscience,*[38] the researchers found that the size of our amygdala increases in times of anxiety and depression. At the same time, the hippocampus, the part of our brain linked to memory and learning, shrinks. Simply put: In stressful environments, our ability to learn and remember things is compromised. Plus, our brain falls back into tried-and-true patterns from the past that ensured its survival then. Considering new or unusual ideas has become virtually impossible.

Burned-Out Circuits: The Propensity to Shortcuts

By the way, if any of you might be thinking life in the cockpit of a megaproject is about power and glory, think again. Delivering major initiatives is not a walk in the park. Aside from working long and arduous hours, project leaders deal with a variety of issues ranging from rumblings of dissatisfied stakeholders to bad publicity in the press, from performance issues to highly complex supply chains, from macro issues such as geopolitical or macroeconomic changes that might impact the project, all the way to micro issues such as their health and well-being. And micro matters just as much as the macro. You cannot lead the project if you suffer burnout.

How does the brain deal with such a plethora of complex inputs? One way is to use mental shortcuts, technically referred to as heuristics. A shortcut might be that when you look out the window in the morning before you go to work and the sky is grey, you take an umbrella. You don't have to analyze whether the umbrella is appropriate every morning. You just take it. Your brain automates the procedure. "This sky looks like it looked the last time I was caught in the rain without an umbrella; better avoid that forever after."

Such shortcuts make for efficiency. They are mental patterns based on previous experience. Think of heuristics as the "Copy-Paste" function in your brain. Usually, these shortcuts work; sometimes, they don't. When they don't, they are called "cognitive biases." A cognitive bias is an error in "seeing": It's when an individual has a distortion in their personal beliefs, a discrepancy between reality and their perception of reality, and this distortion influences their reasoning and decision-making.

When you sit in a meeting with the client, you present the project status, and the client raises an eyebrow, your brain tells you immediately, "Oh boy, here we go again, the client is unhappy, I'll be fired, I must deflect the blame and cover my a__." That's a small example of cognitive bias. These reflexes are based on the past. The brain says, "This is like that other time when I got hurt; I must protect myself the same way." Cognitive biases have an impact on the performance of projects. They can distort all the significant phases of a project, from governance to the business case, from scheduling to cost estimations, and from procurement to project delivery.

Overconfidence or Optimism Bias

Overconfidence results from a false, exaggerated belief in one's own (or sometimes others') ability to produce results. It tends to convince us that we are better than we really are. For example, 73 percent of American drivers consider themselves better-than-average drivers, which is a logical impossibility. This means that at least 23 percent of drivers (the delta between half and those 73 percent) are overconfident. Men are even more full (or is the word "even fuller"?) of themselves: A full 80 percent (four-fifths) of male drivers say they are better-than-average drivers.[39] With some 229 million licensed drivers (not counting an additional estimated 11 percent of drivers without a license) on the roads, at least 52 million of them think they drive better than they really do. The consequences of such optimism bias are dire: In 2019, some 36,000 deaths were through car accidents alone.

Optimism bias is dangerous not only in traffic but also in megaprojects, where success depends on accurate estimates for completing all the works and tasks that make up the project. The researcher and megaproject expert Bent Flyvbjerg found that optimism bias has a systematic, detrimental impact on projects. His extensive research showed how optimism bias, and inaccurate forecasting that stems from it, regularly hamper large-scale infrastructure projects, whether in building rail, bridges, tunnels, or roads.[40]

Flyvbjerg's conclusions: Many project managers cook the books or lie outright to make their projects or companies look better than they really are. Project owners might inflate the benefits of a project while understating the costs.

Daniel Kahneman was the first Nobel laureate for economics who was not an economist. Kahneman is a psychologist who was one of the founding fathers of behavioral economics with his legendary collaborator Amos Tversky. He and other behavioral economists like Richard Thaler and Dan Ariely coined many of the terms below.

According to historians, we are hardly the first generation whose megaprojects suffer from over-optimism—already Alexander the Great did. In the 4th century, before the common era, the ancient king of Macedon proposed building a series of ziggurats. Alexander also wished to make a road from Egypt to the Pillars of Hercules and circumnavigate the African continent. And rumor has it that a plan was drawn up for an entire town to be carved into the side of a mountain, in the shape of a massive Alexandrian hand (of course). None of these grand designs ever materialized.

Then there was Napoleon's Channel Tunnel. By 1802, Napoleon Bonaparte had ordered a feasibility study for a tunnel that would allow him to invade Britain. The engineers found that the megaproject would suffer from a giant difference in sea level and subsequent high costs. Napoleon did not like to hear that and ordered a second report, which ignored and/or falsified the evidence. A toxic mix of prestige, nationalism, imperialism, and megalomania led the French leader, not unlike some current would-be Napoleons, to condone hiding inconvenient facts when they did not suit his strategy.

As a point of fact, there have been many plans for an English-French Channel tunnel, quite a few, before the Chunnel finally came to fruition: It had first been suggested in 1751 by Nicolas Desmaret, then after 1802, it was again designed by French engineer Aimé Thomé de Gamond. Another attempt, this time a joint Franco-British venture presided over by British Member of Parliament Colonel Beaumont, went to the digging stage and was slated to complete by 1880, but was abandoned after one mile of digging because of logistical and political reasons issues plaguing it. Yet another tunnel was proposed in 1919 by British Prime Minister Lloyd George, but the French government swiftly dismissed the idea.

Optimism bias continues to this day. Kahneman revealed a systematic fallacy in planning and decision-making: People underestimate the costs, completion times, and risks of planned actions; in contrast, they overestimate the benefits of the very same actions. This would later be known as "the planning fallacy."[41]

At a conference of project professionals in Asia, we asked for a show of hands for those who had experience writing a business case. Some 20 percent of the participants raised their hands. The next question was telling; "How many of you actively tested the assumptions or tried to "kill the case" by altering the assumptions?"

All hands went down. Not a single hand stayed up.

The problem is that once you make a business case, you tend to fall in love with the case, you tend to be biased in favor of the case, and you tend to ignore the evidence that speaks against the case. (The same can happen when you write a book, but that's another matter, we are being entirely hypothetical here.)

Confirmation Bias: Seeing What We Already Know

If our only biases were over-optimism and negativity dominance, we could manage. But alas, there are a plethora of other distortions. One of them is Confirmation Bias: The human brain registers information and data that confirm its pre-existing beliefs. And it tends to ignore any information that might run contrary to those beliefs.

Did you ever have this experience? You bought a stroller for your child or just became a mother or father, and suddenly the street was full of strollers? Or you bought a new car, and for days after your purchase, you saw lots of those same cars? That's confirmation bias of the innocuous sort. But this bias can easily do damage. For example, when managers interview job candidates, 60 percent of interviewers have made up their mind about the candidate within 15 minutes, and 30 percent of interviewers even within 5 minutes.[42] Worse, when you see a political candidate on TV or the Web, your brain makes a snap judgment before consciously deciding whether to vote for that politician. A 2005 study showed Princeton students photos of candidates from the last three Congressional races. As each pair of candidates came up on the computer screen, students had to judge quickly which candidate looked more competent. It turned out that students picked the actual winner almost 70 percent of the time.

Before we even blink our eyes, our brain has decided whether we want to hire, date, hate, or make friends with the person we just met. And then the brain registers only the evidence that supports its snap judgment.

Since our perception of reality is "systematically biased", as Daniel Kahneman and Amos Tversky put it, this should not come as a surprise: our performance is constrained by our mental models based on our database of past experiences. According to British neuroscientist Richard Gregory, 90 percent of what we see is projected from our memory.[43] Only 10 percent is added by fresh input from our sensory organs. Another neuroscientist, Jeff Hawkins, wrote about a "memory-based model of the brain":

The brain uses vast amounts of memory to create a model of the world. Everything you know and have learned is stored in this model. The brain uses this memory-based model to make continuous predictions of future events (...) To make predictions of novel events, the cortex must form invariant representations. Your brain needs to create and store a model of the world as it is, independent from how you see it under changing circumstances.[44]

At an industry seminar with public-sector clients, we asked the participants: "What do you already know about contractors?" Here are some of the responses:

- "You can't trust them."
- "They will try and cheat you."
- "They always put forward their A team during bidding, but what you get at delivery is the Bs or the Cs."

Sound familiar?

Then we asked the follow-up question: "Imagine that you are now a contractor whose values are "Integrity, Trust, and Performance. How easy would it be for you to work with a client like yourself?" The silence in the room was deafening.

That is confirmation bias at its finest. Once we have formed an opinion on "those contractors" (and believe us, that opinion was developed long before you entered the picture), our brains will register only the evidence, which cements that opinion only further. (See what comes up if we say "lawyers" instead of "contractors." Or "consultants," for that matter. Or "Bill Clinton." The moment you hear "Bill Clinton, what comes up in your brain? Certainly not "great economic boom of the 1990s" or "the Clinton Foundation", but instead, we bet, "Monica Lewinsky".)

This projected memory distorts what is out there and influences how we deal with problems and issues as they arise. It can significantly impact performance, especially in dealing with complex challenges that require innovative ideas depending on the brain's ability to form new neural connections.

Authority Bias: Leader Knows Best

The potent ability of those who are (or are seen to be) in positions of authority to exert influence and control others was demonstrated in a remarkable set of studies performed by Stanley Milgram.[45] In particular, Milgram was interested in understanding how people were willing to be obedient and follow orders, even if the demands meant inflicting harm on other people. The experiment and its results regarding social influence were both telling and alarming. The experiment involved placing ads in newspapers to recruit participants from various backgrounds. Candidates were told that the research goal was to study the effects of punishment on learning. The researchers said that participants would either be teachers or learners, but the process was rigged so that all participants ended up being teachers. Each participant, the "teacher," and the learner (the latter played by an actor) were taken into a room where a screen separated both so the participant couldn't see the learner. The person in authority was played by a third person on the participant's side, providing the instructions.

The core of the experiment involved each participant asking the "learner" a series of questions. Every time the "learner" got a question wrong, the person in authority instructed the participant to "punish" the "learner" by turning a dial and administering electric shocks. (In reality, the "learner" was an actor who simply played pre-recorded screams every time the voltage increased.)

The experiment found, shockingly, that a very high proportion of subjects fully, albeit some of them reluctantly (and some with "nervous laughter"), obeyed the instructions. Had the fake electric shocks been real, they would have killed the learners.

Within the context of project management, we are not saying that all people in authority harbor evil intent or that they instruct people to do harmful things to others. Instead, we are saying that hierarchy has an unintended side-effect: Those below the top of the organization chart tend to seek guidance and affirmation for their actions and take the path of least resistance. In other words, to obey.

Great project leaders create environments where direct reports can openly and freely discuss risks, give bad news, and put issues on the table. By contrast, poor leaders enhance a climate of fear and anxiety. The result is a project where any bad news is filtered heavily as it travels from the field and up the organizational structure. As the saying goes, "Pretending it's not there doesn't mean it's not there."

You might object, "But don't nice guys finish last? Don't you have to be tough to get anything done? Didn't Jack Welch succeed by sacking the 10 percent lowest performers each year? Didn't that climate of fear spur people to perform at their best?"

The answer is, yes, for a while perhaps, but the performance won't last. And it will depend on one strong personality. The management theorist Jim Collins described this in *Good to Great* as "a pattern we found in every unsustained comparison: a spectacular rise under a tyrannical disciplinarian, followed by an equally spectacular decline when the disciplinarian stepped away, leaving behind no enduring culture of discipline."

A typical example was Rubbermaid —the U.S. manufacturer of countless household products such as trashcans, closets, laundry baskets, and step stools- under the helm of Stanley Gault, who was accused of being a tyrant and quipped in response, "Yes, but I'm a sincere tyrant." The result: Rubbermaid rose dramatically under Gault—it beat the market 3.6 to 1, and its stock rose an impressive 25 percent each year—but after he left, it lost 59 percent of its value relative to the market before being bought out by Newell.[46]

Herd Mentality and the Spiral of Negativity

Herd mentality is the tendency to copy and follow blindly what others are doing. At the core of herding is blind faith in trusting the mindsets and cues provided by others. It's the proneness to go with the crowd, influenced by emotion, rather than stepping back, doing an independent analysis, and drawing your conclusions.

Social behavior research by the University of Leeds found that humans have a flocking instinct similar to the collective behavior of other animals such as swarms of bees, flocks of birds, or schools of sharks.[47] In a group experiment, some 200 volunteers were asked to walk around a large hall without talking to each other. A few test subjects were given detailed directions on which route to take. The scientists found that people have a blind tendency to follow other people who appear to know where they're going. The statistics were shocking: It takes only 5 percent of confident-looking people to influence the direction of the other 95 percent. And this phenomenon happens without people being aware of it.

Studies by the University of Exeter[48], in collaboration with Princeton and Sorbonne Universities, also showed how our natural desire to be "part of the crowd" distorts and damages our ability to make the right decision. This can result in groups becoming insensitive and unresponsive to changes in the external environment, as their focus and attention are absorbed into copying one another.

Emory University neuroscientist Gregory Berns calls this phenomenon "the pain of independence,"[49] a situation where taking a stance different from the groupthink results in the activation of the amygdala and initiation of a fear of rejection response.

When projects get into trouble or are distressed, herd mentality combined with negativity bias can generate a deadly, toxic environment in which the project (dys)functions. Walk through the offices of an under-performing project, and you'll likely see people with their heads down, stressed out, frustrated, and fearful that they will not hit significant milestones. Since negativity dominance is firmly anchored to the amygdala, the outcome is a cocktail of anxiety.

Sunk Costs—"Goodbye" Is the Hardest Word

Most of us make a wrong decision that we later come to regret at some point in our lives. Examples could be picking the wrong job, choosing an incompatible life partner, or discovering you have put money into a terrible business idea. Making a single lousy judgment call is one thing, but continuing to invest in it is another. It sounds insane, but that's precisely what many of us do. Consider how many people you know that are holding onto bad jobs, toxic work environments, unhealthy diets, and are stuck in unproductive practices.

Psychologists call this phenomenon the "Sunk Cost Effect," It is a bias that runs rampant in many aspects of our personal and corporate lives, including projects.

Ever had the experience of forcing yourself to watch an awful movie simply because you paid and purchased the ticket. This is the

Sunk Cost Effect at play. Logically, it would make sense to leave the cinema and do something more pleasurable with the remainder of the movie time. After all, the cashier (or, more likely, the streaming service at home) won't give you your money back simply because the "movie sucked." But that's not how we operate. We continue to invest in losing propositions because of what they have already cost us. And this is precisely what we would do with the movie. Upset and irritated, we sit through the final scenes, waiting for the curtains to close, lights to come up, and then race out of the exit.

At the heart of the Sunk Cost Effect are negative vicious cycles associated with loss aversion, its close cousin.

Daniel Kahneman and Amos Tversky conducted an experiment:

- "You are offered a gamble on the toss of a coin.
- "If the coin shows tails, you lose $100.
- "If the coin shows heads, you win $150.
- "Is this gamble attractive? Would you accept it?"[50]

Through many such experiments, they found that "For most people, the fear of losing $100 is more intense than the hope of gaining $150." Hence, they concluded that "'losses loom larger than gains' and that people are risk-averse."

On the surface, Loss Aversion has the illusion of being prudent, cautious, wary, or far-sighted. But without being checked, it can do the opposite. It's one thing being trapped in a 5,10 or 15 US dollar movie, quite different from being imprisoned inside a badly-failing $20 billion project.

According to this theory,[51] a loss is perceived as more severe and more worthy of avoiding than an equivalent gain. A definite gain is preferred to a probable one, and a probable loss is preferred to a definite one. Because we want to avoid definite losses, we look for options and information with certain gains.

Our innate inability to accept failure and cut losses can leave executives, corporations, and projects trapped in the same vicious cycle that gamblers get into, pouring good money after bad, hoping to recover the losses.

The Sunk Cost Effect and Loss Aversion are both nasty. Nasty in the sense that they drain resources, waste money, and steal valuable executive attention. Not to make a political statement here, but: Successive U.S. administrations, both Democratic and Republican, have suffered from the Sunk Cost effect in all wars the United States has fought after World War II. Whether it was Korea, Vietnam, the Gulf Wars, or lately, the debacle in Afghanistan after a 20-year engagement: Once the United States had committed significant resources on the ground, it became virtually impossible to escape the Sunk Cost Effect. The same could be said about the Russian government's 2014 and 2022 invasions into Ukraine.

Bottom Line: Systematically Skewing Reality

The biases we have illustrated above are by no means fully exhaustive. Several sites on the internet have attempted to categorize them systematically. The main point we want to get across is how easily humans can distort, skew or completely ignore reality; and how quickly our biases have us get out of touch with the real drivers of performance.

There is real value in realizing that the human brain does not faithfully reflect reality—and by "real value," we mean financial value, as in dollars and cents. Our brains skew, distort, or completely ignore the incoming data and instead validate an invalid map of reality. Essentially, our cognitive apparatus is built to undermine performance through three distortions:

- We ignore incoming data.
- We skew information processing; and
- We generate false conclusions and take ineffective actions, making the outputs highly variable.

And here is the scary finding by cognitive scientists: Bias and distortions are not the exceptions. They don't happen sometimes. They happen *systematically*. It's not the case that we have a faithful image of reality, and our brains make an occasional blip or error that tricks us. No: Our brains distort reality all the time. The next chapter provides a visual tool to categorize the broad spectrum of biases for practical application.

Radar Alert 9

BUILD AN ADVANTAGE BY MAPPING AND SCANNING

THE COMPLETE TERRAIN.

Chapter 9 > Looking Under the Hood

As Carl Jung put it,

"In each of us there is another whom we do not know."

As Pink Floyd sang,

"There's someone in my head, but it's not me."

— David Eagleman, Neuroscientist

In this chapter, we unite two fundamental ideas of this book: (1) Each project has its unique flight path; and (2) The flight path, and hence success or failure, are impacted by hidden forces in the Black Box. Chapter 3 offered the former; the previous chapter delved into the neuroscience of the latter.

Central to human beings is our ability to perceive, emote, decide and act. And we are not limited to this; we also have a great capacity to put ourselves in another human's shoes. This power is thoroughly hardwired into our system: Studies have shown that we mirror other people's emotions. That's why we cry when we watch movies (and, in Thomas' case, even AT&T commercials). More seriously, we and certain primate species and birds have so-called mirror neurons that allow us to empathize with other people's situations, motives, and actions. For example, in 2005, Marco Iacoboni and others at UCLA reported that mirror neurons could discern whether another person about to pick up a cup of tea from a table planned to drink from that cup or clear it from the table. [52]

In humans, brain activity consistent with mirror neurons is found in:

1. The premotor cortex.
2. Supplementary motor area responsible for the control of movement.
3. Primary somatosensory cortex (responsible for the sense of touch), and
4. The Inferior parietal cortex (helps individuals orient themselves spatially}

In plain English, this means that mirror neurons fire in specific areas across the brain to "mirror" the behavior of the other, as if the observer were him- or herself acting. This is why we get teary-eyed when we see our hero cry in a movie (and why Thomas gets moved even when he watches a commercial for a phone company or a pet shop).

This mind-reading ability is not an exact science nor 100 percent foolproof. But it does allow us to get a glimpse into another person's world. We can picture life through the eyes of another. We can obtain valuable intel: insights into the other's thought processes, an appreciation of their feelings, and a comprehension of the beliefs and perceptions underpinning their behaviors and actions. In short, we can get an idea of "what's going on over there, without physically being there."

Moreover, what happens in our minds and brains, not least our beliefs and biases, directly impacts the success or failure of any endeavor (yes, even megaprojects). We will illustrate this in four mini-case studies that show the dramatic impact of the Black Box on project outcomes. As you review each megaproject, consider how the biases and beliefs "boxed in the project" and shaped its trajectory. Think about the substantial impact they may have had on the hard tangibles such as:

- Product and stakeholder satisfaction
- Program or schedule accuracy
- Precision and containment of costs, and

- Riskiness of the overall project

We have grouped the ever-expanding list of biases into three categories (see Figure 9.1 and Table 9.1 below) to make understanding this mental world of projects easier:

1. "I" = Self-oriented biases that act systematically upon us, both inside and outside a project.
2. "We" = Group and social biases that affect our collaborations, teams, and organizations.
3. "It" = Decision-making biases that condition our ability to evaluate information, assess situations effectively, make high-quality decisions, and perform with cognitive flexibility when circumstances change.

The aim here is not to provide a complete list of biases; we have commented on this in Chapter 8. We want to offer some distinctions that might be useful when you're in the thicket of leading an actual project.

Managing the Portfolio Black Box: Common Biases

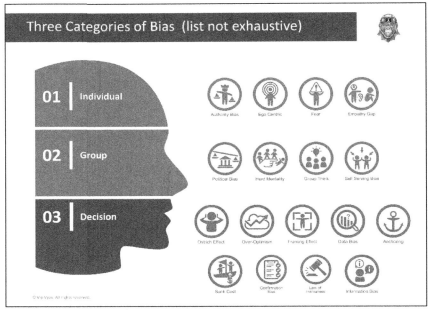

Figure 9.1: Categories of Bias.

Category	Name	Description
Personal "I"	Authority Bias	Overestimating the weight and accuracy of an authority figure's view and being overly influenced by their view
	Empathy Gap	Underestimating the influence of one's own, or others', emotions and feelings in making decisions
	Ego Centric	Relying more heavily on one's own point of view than on others. Thinking that one's influence and importance are more significant than they are
"I"/"We."	Fear	Being distressed by emotion or aroused by impending danger, whether the threat is real or imagined
Social "We"	Political Bias	The tendency to change or "slant" information to promote a political position (or political candidate), making the person or idea seem more attractive
	Self Serving Bias	The inclination to attribute our successes to ourselves and our failures to external factors

Category	Name	Description
Social "We"	Group Think	Suppressing diversity and dissent through the desire for harmony and conformity, resulting in dysfunctional decision making
	Group Think	Conforming our behaviors or beliefs to those of the group to which we belong (also called Group Think)
"It" (Decision Making)	Ostrich Effect	Avoiding information we perceive as potentially unpleasant or inconvenient
	Data Bias	The belief that the data set you have is relevant to the model or application you are working on
	Law of Instrument	Using the same "tool" for every purpose. Often expressed by the saying, "If the only tool you have is a hammer, you treat everything as a nail."
	Anchoring	Relying too heavily on or "anchoring" on the first piece of information seen or heard about a particular topic. Then registering only the evidence that confirms the initial anchor (also called Priming)

Category	Name	Description
"It" (Decision Making)	Confirmation Bias	Interpreting new data and information in a way that confirms our own preconceptions
	Framing Effect	Drawing different conclusions from the same information based on the context in which the data is presented
	Information Bias	Believing that more information will enable better decisions. The data is often irrelevant to the current issue.
	Over-Optimism	Overestimating the probability that desirable (positive) outcomes will happen, coupled with underestimating the likelihood of undesirable (negative) outcomes
	Sunk Cost	Making decisions based on what we have already invested in time, effort, and money, even where future returns are not worth the extra expense.

Table 9.1: The Project Black Box, Common Biases

Thinking About Thinking—The Top 3 Deep Radar Questions

Each mini-case starts with a project brief. We suggest you read this first and then pause to reflect on what you have read. Consider three questions:

1. Were the main challenges on the project mainly individual, social (political), or primarily technical?
2. What beliefs, biases, and perceptions were shaping behaviors on the project?
3. How could project performance have been impacted if the stakeholders had reviewed the Black Box, discussed it, and acted upon it in each phase of its life cycle?

Armed with these lines of inquiry, it's now time to stand up, walk over and step onto the flight decks of four very distinct megaprojects, and examine what happened in their Black Boxes. You may want to bring your pad and pencil.

Radar Alert 10

WATCH FOR HIDDEN BELIEFS, BIASES, AND ATTITUDES

PLAYING OUT ON YOUR PROJECTS.

CHAPTER 10 > MINI-CASE: SYDNEY OPERA

The larger the group, the more toxic,

the more of your beauty as an individual

you have to surrender for the sake of group thought.

— George Carlin

Situated on Bennelong Point in Sydney Harbour, resembling a flotilla of sailboats (or a Tsunami?), the Sydney Opera House is one of the world's most famous landmarks. It is a magnificent building, but its story is bittersweet. Its prestige and status came with a massive price tag and several fiascos.

Planning for a new Sydney Opera House began in the late 1940s when Eugene Goossens, the Director of the NSW State Conservatorium of Music, lobbied for a suitable venue for large theatrical productions.[53] The Sydney Town Hall, the default venue for such performances, was not considered large enough. By 1954, Goossens succeeded in gaining the support of New South Wales Premier Joseph Cahill, who called for designs for a dedicated opera house.

The city launched an international design competition and received 233 entries from architects in 32 countries. The criteria specified a large hall for 3,000 people and a small hall for 1,200, designed for different uses, including full-scale operas, orchestral and choral concerts, mass meetings, lectures, ballet, and other performances.

The winner was announced in 1957: Jørn Utzon, a Danish architect. Legend has it that the Utzon design was rescued from a final cut of 30 "rejects" by the noted Finnish American architect Eero Saarinen. The prize was £5,000 (less than USD 7,000). Utzon visited Sydney in 1957 to help supervise the project. In 1963 his office moved to Sydney's Palm Beach.

Forty years later, in 2003, Utzon received the Pritzker Architecture Prize, architecture's highest honor. The Pritzker Prize citation read: "There is no doubt that the Sydney Opera House is his masterpiece. It is one of the great iconic buildings of the 20th century, an image of great beauty that has become known throughout the world—a symbol for a city, whole country, and continent."

The Politics, Winning Votes, and Major Projects

That was not how people in Sydney saw Utzon at the time. Far from it.

The government had pushed for work to begin early, fearing that funding, or public opinion, might turn against them. However, Utzon had still not completed the final designs. Moreover, he had never personally visited the site before winning the competition. Major structural issues remained unresolved.

By January 1961, work was running 47 weeks behind, mainly because of multiple unexpected difficulties: bad weather, a struggle to divert stormwater, construction beginning before accurate construction drawings had been prepared, and changes to the original contract documents. Work on the podium was finally completed in February 1963. The forced early start led to significant later problems, not least of which was that the podium columns were not strong enough to support the roof structure and had to be re-built.

Restructuring Governance Mid-Flight

In 1965, the state government changed. The new premier of New South Wales, Robert Askin, put the project under the Ministry of Public Works jurisdiction. The ministry's constant criticism of the project's cost and time overruns, and its view that Utzon's designs were impractical, ultimately led to his resignation in 1966.

The original cost and schedule estimates in 1957 projected £3.5 million (US$7 million at the time) and a completion date of Australia Day, 26 January 1963. The project would be completed, in reality, ten years late and 1,357 percent over budget in real terms (more than 13 times the initially quoted cost). Without a doubt, the final product is a fantastic piece of architecture. But did the journey have to be so costly and painful?

Pause. Let's stop for a moment and dig a bit deeper:

- What beliefs and cognitive biases do you detect are shaping the Background?
- What impact does political ego have on the project?
- In what way is fear coloring communication and interactions?

Even if you don't know the whole story, can you already come up with some possible Black-Box factors that might have derailed the megaproject?

In Figure 10.1 and Table 10.1, we take a peek into the Blackbox: the hidden crash factors.

Summary of the Flight Path

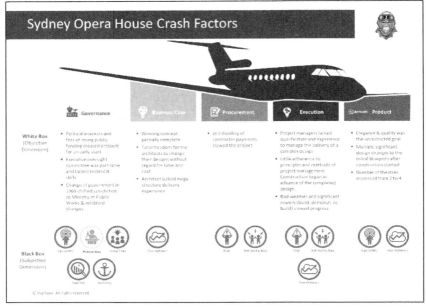

Figure 10.1: Sydney Opera House, Crash Factors

Sydney Opera House Crash Factors

Project Phase	Crash Factors	Beliefs & Biases	Impact
Governance	Government pushed to start the project early for fear of adverse public opinion and loss of funding. The Executive Committee overseeing the project lacked technical skills. Change of government (1965) placed the project under a different jurisdiction (Ministry of Public Works). The government became an obstacle by inhibiting changes.	Political Bias Ego. Fear. Self-Serving interests.	Schedule Risk↑ Cost Risk ↑
Business Case	Project Goal: Quality was the unrestricted goal. No limits on costs or schedule.	Optimism Bias.	Wide Cost Uncertainty
Deliverables	Frequent scope changes lead to a fluid, complex design. Design changes after construction started included, e.g., increasing the number of theaters from 2 to 4.	Impact of an incomplete business case lacking a rigorous understanding of end-user requirements. & Optimism Bias	Product Uncertainty ↑ Product Benefit ↓

Procurement	Winning architectural concept was partially complete. Project cost estimates based on preliminary and incomplete design drawings. The withholding of payments to contractors during execution slowed down the project.	Optimism Bias. Fear was used as a weapon to force contractors. A lack of Empathy and understanding of the delivery chain predicament.	Collaboration Risk ↑ Cost Risk ↑ Schedule Risk ↑
Execution	Architect walked off the project without leaving behind the designs. No formal project manager was in place & principles and project management methods were not adhered to. Interactions between Utzon and Arup dictated delivery. Pressure to start construction in advance of complete designs leading to significant rework - build/demolish/rebuild. Bad weather slowed progress	Project execution was a blind spot. Architects' role as the project manager may have undermined objectivity regarding project performance.	Accountability Uncertainty ↑ Capability Risk ↑ Abortive Works ↑ Costs ↑ Schedule Risk ↑

Table 10.1: Sydney Opera House, Crash Factors Along the Project Cycle

Opening the Project Blackbox: Yes-Men

Sydney Opera is a powerful illustration of how a compounding mix of politics, over-optimism, and a fear-driven culture can undermine governance and project delivery. From the outset, the Sydney Opera

House story is that of politicians eager to fast-track and prioritize their interests and glory over the interests of the public and the treasury.

Not only that. Imagine being invited to a project review meeting with New South Wales Premier Joe Cahill in attendance. Here is what you need to know about Joe. Of Irish descent, Joe Cahill was an active trade unionist. Here are some words used by those around him to describe his personality: "tough," "bully," "activist," "controller," "manipulator," and "agitator."[54]

Given these descriptions, how willing do you think those around him were to stand up to him and his followers and take exception? What "unchecked risks" could have been generated in those meetings with a formidable leader at the head? That's herd mentality at its finest.

Risks Internally Generated

Risk management—Identifying, assessing, and controlling project threats—is ubiquitous in megaprojects and large-scale change initiatives. The built-in underlying belief of conventional risk management is the need to identify all external factors that could jeopardize a project. Central to this practice is the view of relating to risks as an "external phenomenon."

Our experience of consulting hundreds of projects, project sponsors, and delivery teams rarely considers risks as an internally generated process. Their natural inclination is to "look outward." The Flight Path framework aims to counterbalance this bias, enhance project intelligence, and achieve more reliability. But first, we will look at another mini-case, the mother of all megaprojects: the Joint Strike Fighter.

Radar Alert 11

EMBEDDING YOUR EGO INTO YOUR PROJECT

CAN MAKE YOU BLIND TO REALITY.

CHAPTER 11 > MINI-CASE: F-35 JOINT STRIKE FIGHTER

The mark of an intelligent person is the ability

to hold competing facts/ideas and still function.

— F. Scott Fitzgerald

With lifetime costs exceeding $1.7 trillion, Lockheed Martin's F-35 Joint Strike Fighter takes pole position as the most expensive weapons system in U.S. military history,[55] and a project cost overrun of $160 billion to match.

To put the costs in perspective, we can compare them to the United States' trade deficit with China, which in 2020 stood at $916 billion.[56] The Joint Strike Fighter has cost close to double—so far. This is serious money. The cost of the Joint Strike Fighter exceeds the GDPs of Norway, United Arab Emirates, South Africa, and Hong Kong—put together. The cost overrun alone equals the total government budget of Finland.

The U.S. Department of Defense invested all this money in a simple concept. Think of the F-35 as the Swiss army knife of aerial warfare, a jack-of-all-trades. But unfortunately, it is proving to be a master of none. True, the F-35 can stealth-spy, drop bombs and engage in dogfights. But, it underperforms on all these tasks compared to older dedicated aircraft. The much older Boeing B-52 Stratofortress bomber, for instance, can fly much further and carry larger payloads. And when it came to ground support missions, such as attacking armored vehicles or providing close air support for ground forces, the F-35 is outperformed by 1970's A-10 Thunderbolt, over 40 years its senior. In an article for *The Hill*, military expert Sean McFate wrote: "Astonishingly, the F-35 cannot dogfight, the crux of any fighter jet. According to test pilots, the F-35 is 'substantially inferior' to the 40-year-old F-15 fighter jet in mock air battles."

The F-35 could not turn or climb fast enough to hit an enemy plane or dodge enemy gunfire. Similarly, the F-35 struggled to get a clean shot at a 1980s-vintage F-16. The older aircraft easily maneuvered behind the F-35 for a clear shot, even sneaking up on the "stealth" jet. Despite the F-35s vaunted abilities, it was blown out of the sky in multiple tests. [57]

Defense experts were not the only ones to express concerns about the F-35's shortcomings. The powerful chairman of the House Armed Services Committee, Rep. Adam Smith, described the Joint Strike Fighter program with his typically acid tongue as a "rathole" for taxpayers. At a Brookings Institution session, the Congressman added: "What does the F-35 give us? Is there a way to cut our losses? Is there a way to not keep spending so much money for such a low capability? Because the sustainment costs are brutal." [58]

Many have described the F-35 as a public relations nightmare. Over the years, the media has had a field day (many field days) with the failed defense program: "The US Air Force Quietly Admits the F-35 is a Failure." "Lockheed F-35's Tally of Flaws." "The Hidden Troubles of the F-35." "F-35 Tells Everything that's Broken in the Pentagon." And finally, "Stop Throwing Money Down F-35 'Rathole'."

History of a $1.7 Trillion Failure

So how did we get here? How did leaders of so many countries end up pouring billions of dollars into a poorly performing defense system whose usefulness in actual missions is, to put it mildly, questionable? What fundamental beliefs and biases were running during the early concept and investment phases? And how did these impact the later stages of the program, such that the multi-mission aircraft program ultimately failed to deliver on expectations?

To understand the unraveling of the fiasco, let's start at the beginning. The Pentagon's rationale was to develop a single Common Affordable Lightweight Fighter (CALF). This major upgrade would replace a wide range of existing fighter, strike and ground attack aircraft for the United States and its allies, including Australia, Canada, Denmark, Italy, the Netherlands, Norway, Turkey, and the United Kingdom. In the defense industry at the time, the idea of a one-size-fits-all warplane was not new. In the early 1990s, following the collapse of the Soviet Union and the end of the Cold War, when the United States suddenly stood as the sole hegemonic superpower, the U.S. Air Force built a program called Multi-Role Fighter. Not to be outdone, the U.S. Navy had its own program called Advanced Fighter-Attack. But fantasies abounded about the "end of history" and the end of all warfare, and the political will was no longer there to spend a lot of money on military hardware. These budget pressures led to the cancellation of both programs. But in 1993, a new program emerged: the Joint Advanced Strike Technology. Boeing and Lockheed Martin competed for the development program via their respective X-32 and X-35 designs. Lockheed won, and the final design was based on its X-35.

To cut a long (and painful) story short, the program design process could have considered the less-than-stellar results from Germany's World War II Junker-Ju18 or the Panavia Tornado program in the 1970s, both of which had experienced severe technological challenges. But despite the ready availability of this historical data, there was a fundamental belief among Pentagon officials that the speed of technological advancements would somehow magically overcome the shortcomings of previous similar endeavors.

Another issue was even more fundamental than ignoring history. At the heart of the challenges facing the Joint Strike Fighter Program were the conflicting requirements of its three key clients: the army, the navy, and the air force. True, all three organizations are part of the United States (or any country's) defense strategy. Yet the nature of their work is radically different. What matters most to the Marines, for example, is the need for Short Take-Off and Vertical Landing (STOVL). This need contrasts with the Air Force's long-range missions and air-to-air combat requirements. Not to mention the Navy, for which the ability to land on a fast-moving destroyer is of paramount importance.

Pause. Let's stop here for a moment and ask ourselves two questions:

- Suppose you had already spent $500 billion, and the results were not looking good. How willing would you be to shelve the program? (Hint: Remember the Sunk Cost section in Chapter 6 above?)
- Even without knowing much about the Joint Strike Fighter: Which critical assumptions did the success of this defense program depend upon?

Summary of the Flight Path

When you're engaged in a highly complex endeavor like the Joint Strike Fighter, you can easily get caught up in the multi-dimensional complexity of the program's structure, let alone the blizzard of thousands upon thousands—no, millions—of activities by a myriad of stakeholders: government agencies and politicians in the United States and eight other countries, generals and other officers, test pilots, companies, engineers, project managers, workers, soldiers and so forth.

In the heat of the battle, It is worth stepping back from the details and getting an Air Traffic Controller's view of the unfolding Flight Path. When you're deeply immersed in the day-to-day fire-fighting of project problems, such an aerial perspective, as it were, might reveal aspects of the project that are invisible on the ground.

So just like with the Sydney Opera House in the previous chapter, we have again mapped onto the canvas the Flight Path of the Joint Strike Fighter and some of the key events that transpired. Then, in the second table, we sought to reveal the beliefs and biases that might have jeopardized this mother of all megaprojects.

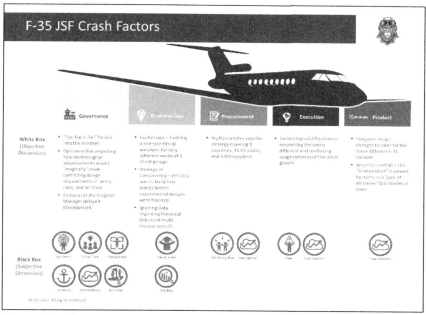

Figure 11.1: The Joint Strike Fighter: Crash Factors

The Joint Strike Fighter Crash Factors

Project Phase	Crash Factors	Beliefs & Biases	Impact
Governance	Removal of JSR Program Manager delayed development.	Anchoring on a dubious idea Framing Effect. Sunk Cost Bias.	Leadership Uncertainty ↑ Team Stability↓ Loss of Knowledge Schedule Risk ↑ Cost Risk↑
Business Case	Concurrency: the idea of building the first planes before finalizing the experimental designs. Ignoring the Data: Disregarding historical failures of multi-mission aircraft.	Confirmation Bias. Optimism Bias: the blind belief that technological advancement would solve all problems. Faulty logic: building a one-size-all warplane could meet the very different needs of 3 client groups (air force, navy, and marines (land force).	Product Risk ↑ Financial Risk ↑
Deliverables	Frequent Design Changes to meet the	Consequence of anchoring	Product Uncertainty ↑

	differing needs of the three F-35 variants (the final product). Benefit Shortfall - A jack-of-all-trades but a master of none?	on a potentially faulty business case	Product Benefit ↓
Procurement	Highly complex supplier strategy covered nine countries, 45 US states, and 1300 suppliers. Very complex interfaces. A mental "Too big to fail" lock-in.	Supplier diversification bias creates both upside and downside	Interface Complexity ↑ Coordination & Integration Risk↑
Execution	Significant technology difficulties in reconciling the vastly different usage pattern needs of the air force, navy, and marines.		Delivery Complexity↑ Abortive Works ↑ Costs ↑ Schedule Risk ↑

Table 11.1: The Joint Strike Fighter, Crash Factors, Beliefs & Biases, Impact

Opening the Project Black Box

The Joint Strike Fighter is an example of faulty logic and optimism bias on steroids. The program is still underway, and there is always the chance of a new "magical solution" that could solve its problems, but this seems unlikely as of this writing.

At the center of the technical and execution challenges were three client groups (military, army, and navy) with divergent—and sometimes polar opposite—requirements. That could have been solved. But other factors were at work in the background.

First, despite this significant hurdle, the Joint Strike Fighter project had a powerful Framing Effect skewing group thinking. There was a fundamental, probably sub-conscious, and never substantiated belief that new emerging technology would somehow redress the shortcomings of the previous similar failed endeavors.

How something (anything) is framed can shape our confidence that it will bring either gain or loss. This is why we find it attractive when the positive features of an option are highlighted instead of the negative ones. And this is precisely how the U.S. Air Force Chief of Staff General Charles "CQ" Brown[59] framed some of the program's shortcomings by comparing the F-35 to a high-end Ferrari, which would only be used in high-end combat.

Second, the Joint Strike Fighter decision-making process is an example of an Epistemic Bubble. By its very nature, a bubble has an "inside" and an "outside." When inside an Epistemic Bubble, a team can fail to expose itself to the complete information needed for effective decision making. This can be done by either neglecting, ignoring, or diminishing the importance of specific data, sometimes unintentionally leading the entire organization or team astray and at the effect of systematic blind spots.

Third and finally, the Joint Strike Fighter is a prime example of the Sunk Cost Effect. With so many nations having (actively) signed up and invested astronomical sums, the megaproject was (and is still) considered "too-big-to-fail." And given this belief, rather than stop the project, the United States and eight other countries doubled down on their financial contributions. The jury is still out on whether (or not) the product will serve the intended purpose effectively.

The Sunk Cost Effect has been studied extensively by psychologists. Interesting findings include Marijke van Putten and her colleagues, who distinguished individuals with a "state orientation" from those with an "action orientation."[60] People with a state orientation struggle to let go of past events and are more prone to exhibit the sunk-cost effect. In contrast, action-oriented people are relatively untroubled by past events. They can let go of what happened and more easily focus on what's next.

But Wait—Are We Missing Something?

Is the F-35 just simply a sinkhole for taxpayers' money, or is there a silver lining to this complex program? Are there merits that we are failing to consider?

The previous description of the F-35 may make this program sound like an unmitigated disaster. As of this writing, the final product is still being fine-tuned and modified. The question is, *"Does our yardstick of success need to be refined with the addition of extra dimensions?"* In hindsight, very few would argue the immense value of Sydney Opera as a global landmark and the associated benefits for the economy of Australia. In a similar vein, could the F-35 be considered a "value creator" standing in the future looking backward?.

The F-35's success may not lie within the Cost-Schedule-Scope triangle but in the cooperation of the nine nations (U.S, U.K, Italy, the Netherlands, Turkey, Australia, Norway, Denmark, and Canada) in creating a defense system that some regard as the centerpiece of 21st-century global security.

Then there are also other political and economic dimensions to consider. For instance, in 2019, a group of Republican senators, including Sens. Marco Rubio of Florida and Ted Cruz of Texas, influenced and urged the Pentagon to consider expanding the program via additional foreign military sales and creating a stronger coalition of allies.

The Swiss Federal Council justified and defended the purchase of 36 F-35As to replace its aging fleets of F-5 Tigers and F/A-18 Hornets: "The F-35A includes entirely new, extremely powerful, and comprehensively networked systems for protecting and monitoring airspace," the federal council said.

Considered by many as a "Program Too Big to Fail," "political purchases" of the aircraft, coupled with the economic benefits of boosting job creation, may provide a compelling narrative for the original proponents of the idea.

Radar Alert 12

FAILURE IS WIDESPREAD BUT NOT INEVITABLE.

MEGAPROJECTS CAN WEATHER THE STORM

AND TOUCH DOWN SAFELY.

Chapter 12 > Mini-Case: Guggenheim Bilbao

Mastery is a process, a journey.

You have got to love the journey—

the sense of not knowing, the sense of learning.

The Path of Mastery lives in the plateaus, not in the peaks.

You have got to love the practices.

Mastery lives in the practicing.

— George Leonard

Not all megaprojects crash. "This is not to say," megaproject researcher Bent Flyvbjerg put it, "that there are no projects that were built on budget and on time and delivered the promised benefits." Not every project suffers the fate of the Pentagon's spy satellite program ($4 billion costs overrun), the International Space Station (over $5 billion overrun), or the Boston Big Dig ($11 billion overrun, by 275%). There are success stories too. We examined them in search of vital success characteristics and best practices.

Spoiler alert: The dominant success factors we found are not technical. They are not financial. They are not just based on better time management or higher efficiency, or more stringent control. They invariably lie in the human component of megaprojects: vision and leadership, mindset and culture, communication, and collaboration.

"A Rare Breed of Project"

When Frank Gehry first met with the President of Spain's Basque Region, several top officials attended, including the Mayor of Bilbao, the Ministers of Commerce, Culture, and Education, and the Director of the Guggenheim Museum. Their question was simple: "Mr. Gehry, we need the Sydney Opera House. Our town is dying."

As he recalls it, Gehry's response was, "Where's the nearest exit? I'll do my best, but I can't guarantee anything." And he delivered. The Guggenheim Bilbao is a spectacular building. And it should be. It is an excellent example of designing a successful flight path, in total contrast to the overshoots and budget explosions of Sydney Opera.

The megaproject's success speaks for itself. Guggenheim Bilbao created 3,800 new jobs and boosted tourism revenues: From 2010 to 2017, museum visitors grew by over 38 percent. Talk of the "Bilbao effect" makes Gehry cringe—"it's bullshit," he said—but the Bilbao Guggenheim has become a model of best practices worldwide for how megaprojects can succeed. In Bent Flyvbjerg's words, "The Guggenheim Museum Bilbao is an example of the rare breed of project."

We agree. From a project management perspective, the Guggenheim Bilbao excelled in many ways. A generation after its inauguration, the Bilbao Guggenheim stands as a shining model of what's possible when megaprojects go right. And it would make Frank Gehry world-famous (he was the only architect to appear on *The Simpsons*—the ultimate stamp of approval).

How could the Guggenheim turn out so successfully? After all, when the project began, conditions were far from ideal. Bilbao's economy had been severely depressed in the 1980s. Ship construction and steel industries had declined throughout the decade because of low-wage competition from Eastern Europe and Asia.

One month before the opening, Frank Gehry visited the Bilbao Guggenheim, and he was in for a shock. "I came over the hill and saw it shining there. I thought, 'What the fuck have I done to these people?'"[61] Gehry would say later that the success of this megaproject took him by surprise.

Despite Gehry's protestations, the Bilbao Guggenheim is far from surprising to us. Nor is it a miracle. It is the logical result of performance-focused leadership and effective practices.

Summary of the Flight Path

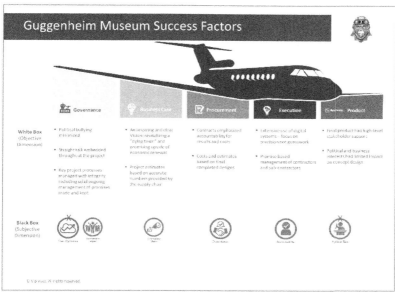

Figure 12.1. Guggenheim Bilbao Success Factors.

Guggenheim Bilbao Success Factors

Project Phase	Key Success Factors	Beliefs & Biases	Impact
Governance	Key project processes managed with rigor and integrity. Use of an effective project control system.	A powerful vision.	Cost and project integrity leadership.
Business Case	Clarity of vision and economic benefits. Consideration of the interests of all stakeholders (in this case, revitalization of a dying town and economic renewal). Project cost estimates based on final completed drawings.	Project estimates based on the reality of the market.	Inspiring project vision. Business Case incorporated realistic cost estimates.
Final Deliverable	A protected design— effective management of political and business interests that could have radically changed	Design protected from political and business interference.	Design integrity maintained. Changes and scope creep avoided.

	the design.		
Procurement	Contracting based on accountability and ownership of estimates.	People accountable for translating their words into deeds.	Supply chain setup for commitment and accountability.
Delivery	Extensive use of digital systems. A focus on precision and reality, not guesswork. Promise-based management of contractors and sub-contractors.	People are central to project success Technology was used effectively to create a delivery advantage.	A high-performance environment created. Technology benefits leveraged.

Table 12.1 The Guggenheim Bilbao, Key Success Factors Along the Flight Path

Key Capabilities Along the Flight Path

Given the sheer overwhelming numbers of so many megaproject failures, something unique must have taken place to account for the success of the Guggenheim Bilbao. What was that? We have distilled ten best practices.

Clarity of Vision. The government cast out an anchor in the future, an inspiring aspiration, nothing less than the reinvention and revival of a whole city. Its commitment was not just to a building or museum but also to enhancing the local economy. The Guggenheim Museum Bilbao was only one but a key component in the government's Revitalization Plan for metropolitan Bilbao, transforming the city into a service metropolis.

Robust Business Case. Juan Ignacio Vidarte, the general director of the Guggenheim Bilbao, whose involvement dates back to the project's conception in the early 1990s, says that the Guggenheim Bilbao was "a transformational project," a catalyst for a broader plan to turn around an industrial city in decline and afflicted by Basque separatist terrorism.

Reality-Based Estimates. A key feature of establishing an effective project control system was Gehry's insistence on cost and schedule estimates based on final completed drawings.

Solid Governance. One thing that stands out was Gehry's intentional focus on creating a high-performance project environment: one where political and business interests would not bully, interfere with, or railroad the design process but were integrated into the process from the start.

The Human Element. An account of the project provided by Gehry in the summer issue of the *Harvard Design Magazine* validates the importance of the human dimension.[62]

Straight Talk. All of this was done through tough conversations (when necessary). To get his point across, as we have seen above, Gehry himself was famous for his love of four-letter words. His leadership by example created a climate of straight talk that made it okay to stop beating about the bush, address issues head-on, and state your point clearly.

Digital Technology. The extensive use of digital systems and data reduced the guesswork that is, alas, common to many megaprojects. This fundamental prerequisite meant that contractors and vendors could bid based on being provided with high-quality data, resulting in more accurate cost submissions and keeping in check any optimism bias.

Contracting for Accountability. Armed with an array of software design visualization tools and rapid prototyping tools to produce physical models provided Gehry's team with the "upper hand" in negotiating favorable contracts. Ones where the schedule and costs could be locked in with great certainty.

Intentionality[63]. The Guggenheim Bilbao was to be a driver of economic renewal, an "agent of economic development" that would appeal to a "universal audience," create a "positive image," and "reinforce self-esteem." All of which, thanks to the single-minded obsession of Vidarte, Gehry, and their government partners, it pretty much did.

Integrity. Gehry's stature and willingness to manage each project phase rigorously were at the heart of a successful flight path. The project was marked by an air of seriousness, with Vidarte and Gehry honoring their word vis-à-vis all partners. They also held all contractors accountable for the schedule, cost estimates, and promises. In this project, "talk was not cheap."

The Guggenheim project fulfilled its original objectives with precision based on these ten best practices. It has been rewarded with a steady stream of a million visitors each year—the 20 millionth visitor came through its doors shortly before the museum's 20th birthday.

Don't Copy This

But copycats, beware: You cannot simply replicate the "stuff." For example, Guggenheims planned for Rio de Janeiro, Las Vegas, Guadalajara, Taichung, lower Manhattan, and Abu Dhabi have mostly failed to materialize or stuttered if they did. And where the copycats failed, they copied the content, not the context.

Psychologists call this the Fundamental Attribution Error. Lee Ross introduced the idea in his 1977 paper "The Intuitive Psychologist and His Shortcomings." He argued that much social misunderstanding is caused by a general tendency to attribute human behavior to personality rather than external circumstances.[64]

Ross called that phenomenon "the fundamental attribution error." Ross demonstrated with an experiment: He devised a game in which Stanford undergraduates drew cards that assigned them the roles of quizmaster or contestant. The quizzer was asked to create difficult trivia questions and pose them to the contestant, who invariably struggled to answer. Other students observed.

After the game, observers said they considered the quizmaster exceptionally knowledgeable and the contestant notably ignorant. That was a fundamental attribution error. Behavior caused by randomly assigned social roles struck those involved as arising instead from intrinsic character traits.[65]

The media (and we) tend to do this whenever someone is successful: We ascribe to leaders and organizations fundamental traits that made them successful. We say, "Richard Branson became a billionaire because he is x and y and z." Millions of other people have had the same x and y and z character traits but haven't been successful. And then there are those other billionaires who have lacked Branson's character traits but have flourished nevertheless.

Or take these Core Values in a company's 1998 Annual Report.

- RESPECT: We treat others as we would like to be treated ourselves. We do not tolerate abusive or disrespectful treatment. Ruthlessness, callousness, and arrogance don't belong here.
- INTEGRITY: We work with customers and prospects openly, honestly, and sincerely. When we say we will do something, we will do it; when we say we cannot or will not do something, we won't do it.
- COMMUNICATION: We have an obligation to communicate. Here, we take the time to talk with one another... and to listen. We believe that information is meant to move and that information moves people.
- EXCELLENCE: We are satisfied with nothing less than the very best in everything we do. We will continue to raise the bar for everyone. The great fun here will be for all of us to discover just how good we can really be.

In the company's heyday, these beautiful core values (character traits at the organizational level) were lauded throughout the 1990s as exemplary by *Fortune* magazine and *The New York Times*.

Did you guess the company? It was Enron.

It's an exquisite example of a fundamental attribution level: We see a company's success and get blinded by the aura of success. We (and the media, management journals, management gurus, and then, of course, the public) attribute good things to that company.

Malcolm Gladwell popularized the fundamental attribution error in *The Tipping Point*: "The mistake we make in thinking of character as something unified and all-encompassing is very similar to a kind of blind spot in the way we process information. Psychologists call this tendency the Fundamental Attribution Error (FAE), which is a fancy way of saying that when it comes to interpreting other people's behavior, human beings invariably make the mistake of overestimating the importance of fundamental character traits and underestimating the importance of situation and context."

Copying Toyota

The operative word here is the last one: context. In the 1980s, carmakers worldwide saw the total quality management (TQM) and continuous improvement at Toyota, the world leader in automobile manufacturing, and sought to emulate it. Toyota even invited them to visit its plants and see how it produced its cars. But many of them had their eye on the wrong ball. They observed Toyota's factory-floor processes and copied them. They installed pull-cords to stop the assembly line if a worker noticed a defect. They set up just-in-time delivery systems. They mandated statistical process control charts. Yet even today, decades later, US carmakers, for the most part, still lag behind Toyota's productivity (the number of hours it takes to assemble a car) and may lag in quality and design features too. Why? Because of an error that happens often (for example, when retailers copied Nordstrom or airlines copied Southwest): People imitated the visible, obvious, and usually least relevant practices.[66] They copy the content, the processes. But Toyota's secret is not a set of techniques.

So, what is it? Toyota's secret sauce is a mindset of TQM and continuous improvement, plus the company's relationship with its workers that has enabled it to tap their deep knowledge. "We have been benchmarking the wrong things," one executive said. "Instead of copying what others *do*, we have to copy how they *think*."[67]

In our view, it goes even deeper than that. We call it context. Context gives how people think. It includes the environment in which the company operates. It consists of the cultural DNA shaped by the founders' values, epitomized by its heroes, rejected by its outcasts, and shaped by defining moments in its history behind Toyota's processes. It includes the totality of background conversations the company's stakeholders (customers, investors, managers, workers) have about Toyota, the industry, and even themselves.

The same is true for megaprojects. That's why the Guggenheim Abu Dhabi and other well-meaning imitations did not take off. You have got to nail the context.

Radar Alert 13

SOMETIMES IT TAKES

TRANSFORMING AN ENTIRE ECOSYSTEM

FOR WORLD-CLASS SUCCESS.

Chapter 13 > Mini-Case—
Leadership in Action at
Shell Pearl GTL

Never doubt that a small group of thoughtful,

committed citizens can change the world.

Indeed, it is the only thing that ever has.

— Margaret Mead

Located in the scorching heat of Qatar's deserts, 80 kilometers north of the country's capital Doha, Shell's Pearl Gas to Liquids ("GTL") is the world's largest plant, turning natural gas into cleaner-burning fuels and lubricants. Pearl GTL is a monster, the largest construction project in the history of oil and gas. With an $18-19 billion development cost, the project deployed 52,000 workers from 59 different nations, with at least as many languages and dialects. Given the project's location, complexity, and sheer scale, an entire town named "Pearl Village" was built to accommodate the workers,

World-Class Performance in Tough Conditions

Safety is a primary concern in the oil and gas and other capital-intensive industries. The work involves performing tasks at dangerous heights, maneuvering multi-ton equipment, or working with high-pressure poisonous gasses. One bad call could cost your life and often the lives of thousands of people.

Also, with summer temperatures exceeding 40°C (over 100°F), the project had to orchestrate hundreds of simultaneous activities in stifling conditions.

In the context of safety, Lost Time Injury (LTI) is a measure of safety performance. A Lost Time Injury happens when an employee gets injured enough to be unable to work the next workday. It's a key metric in project performance: the number of LTI is divided by the total number of hours worked over a given accounting period.

Despite the challenges, Pearl GTL broke all safety records by delivering over 50 million person-hours of work without a single LTI. How? And what were the best practices we can distill from this significant achievement that would not usually have happened?

The Challenge: Scaling Authentic Commitment

A central challenge, perhaps the biggest, is the mindset regarding safety. Enter most construction sites, and you will see a splash of slogans such as *"Safety Is Our #1 Priority"* or *"Safety First."* On the surface, these messages seem to express concern and care. But they don't necessarily echo or reflect workers' daily experiences in many parts of the world. The actual picture imprinted on their minds is quite different. And, when the project comes under pressure to hit looming deadlines, the gap between declared policies and reality only widens. When top management becomes riddled with anxiety and a sense of panic creeps in, a change occurs. Instead of the message, "We care about you, and we appreciate your hard work," a darker mood takes over and clouds the environment. What workers are likely to hear, if not explicitly, then at least between the lines, often sounds more like this:

- "This isn't good enough."
- "I don't want excuses. I want results."

- "You're not gonna tell me you can't get this done, are you?"
- "You better do this or else…."
- "If you can't deliver X by time Y, we'll financially take you to the 'cleaners.'"

People aren't stupid. Such repeated mantras clearly communicate what a corporation cares about most. Whatever it is, clearly, "Safety is not our #1 priority." With the mask removed, the message becomes this: "We expect you to work longer, harder and faster in service of our objectives." There is nothing wrong with these expectations per se; indeed, many businesses operate this way. But within the context of safety, especially in some remote regions around the world, there is an added unspoken sub-text: "We don't care if you have to take a few safety shortcuts or turn a blind eye to safety procedures. What matters is the result. Whatever it takes."

It's no wonder that when the CEO stands on the annual podium and gives the "importance of safety" message, much of the audience fights the urge to roll their eyes. The remainder of the speech is heard through a filter of contempt or, worse, resignation. The tape that likely plays in people's heads goes something like this: "Uhuh, the usual corporate PR bla-bla," or "Here we go again."

Against this backdrop of industry beliefs and biases, what Shell Pearl GTL achieved is astonishing. To have any chance of success, the leadership team had to tackle head-on—and overcome—the deep-rooted skepticism and cynicism rife in the industry.

Designing a Safe Flight Path

The philosophy adopted by Andy Brown, then Executive VP of Pearl GTL, and his leadership team, to avert this rabbit hole of resignation was simple: "If you take care of the people, they will take care of the project." This fundamental principle sounds like a no-brainer, but it heralded a shift in the corporate culture. Rick Bair, the senior partner of JMJ, who worked on the project, recalls: "Brown and the other leaders believed that if every person, every day, was treated with dignity, care and respect, they would return that commitment to the project. And they did."

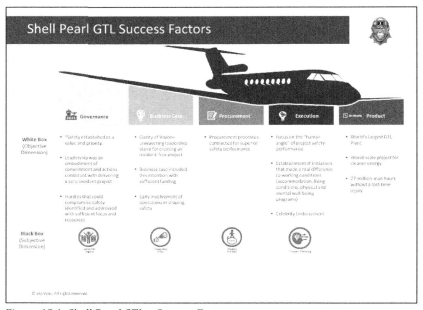

Figure 13.1: Shell Pearl GTL—Success Factors

Governance: Mindsets, Attitudes, Culture

Central to Pearl GTL's success was a profound mind shift across the project. It was a movement toward personal ownership of a new culture: an incident-free safe construction and operations environment.

In the heavy industry sector, most companies say they put "Safety First," but in practice, they do not. In practice, the reality on construction sites, where the rubber meets the road, is vastly different. The everyday experience of many supervisors and workers is of being shouted at and pressured to hit production targets and milestones. As a result, many supervisors and workers hear "Safety First" as political lip-service, a slogan the executives must say to be socially acceptable in the eyes of the media, the public, and the regulators.

This was not the case on Pearl GTL. The project management team worked systematically to build a culture of performance coupled with safety. And since you can easily be blind to your cultural blind spots, they brought in coaches from outside the system. Bair recounts, "When we met with Andy Brown and the senior leadership team to understand their intent for the project, their unequivocal commitment was that Pearl GTL achieve a breakthrough in safety and to do whatever it took." The belief was "cause a breakthrough in safety and the other project metrics would follow." To achieve this objective, Brown and the Shell leadership team focused intensely on revealing, understanding, and addressing the Black Box of this enormous project. As Brown put it, "It was vitally important we created a single purpose to bind people across all cultures."

The benefit of ... an external body if you like, coming in to work with us in this field, and you could call it organization development, behavioral change management, personal transformation," said Julian Johnson, an organizational effectiveness consultant who worked on the megaproject, "the real benefit of that, in a young organization like ours [is] keeping us honest, constantly challenging around these particular themes, I think is a great value, particularly in a young organization like ours, but in any organization, and particularly when you're trying to set the culture from the beginning. You know, Qatar Shell has been, you know, it's only ten years old. [68]

But this safety culture must not live merely on some sleek PowerPoint slides. It had to be embedded across the organization. It had to permeate the megaproject in all its facets. Take, for example, leading by example. How often have you heard the phrase "He/she is full of hot air" (or you can imagine the not-so-polite expression)? People have an acute radar for discerning whether a conversation is authentic or fake. We ask ourselves, "Is this person being genuine, or are they simply talking BS (which, of course, stands for Baloney Sandwich)?" Given the significant life-and-death safety risks facing Pearl GTL, it was imperative that the leadership team "walk the talk" to engage a vast group of experienced and skeptical managers and workers, those who faced the safety hazards every day. "And they did; without the depth of commitment and leadership demonstrated by the many contractors, the spectacular incident and injury-free results on a project of this size and complexity would not have been possible," Bair added. Sheikh Thani Al-Thani, the Deputy General Manager of Pearl GTL, echoed the sentiment when he said, "We have hit many milestones...everybody is committed to completing this project strong."

Procurement

But this commitment to safety could not limit itself to the workforce. It had to permeate all of Pearl's, and ultimately Shell's, activities. Contracts with suppliers, for example, placed safety center-stage, including clauses emphasizing the importance of safety leadership and accountability in creating an incident-free safety environment. These clauses were distinct (and went beyond) the traditional punishment-based approach commonly found in the industry, focusing on safety procedures and process breaches.

Network of Conversations

The megaproject leaders did not stop there. They actively promoted and sustained a network of conversations to highlight and resolve any safety risks with immediate impact. The problem was that many people—especially people from different cultural backgrounds who are not used to working together—often naturally resist correcting each other, worried about encroaching on another organization's territory or stepping on each other's toes. The Pearl GTL leaders overcame this with a simple change: They asked all workers to express their commitment to being safe by wearing Personal Protection Vests saying, "Talk with me about Safety" printed on them. This symbolic statement allowed individuals, regardless of their position, company, or nationality, to intervene with each other in the service of a safe work site. A tiny, targeted rule can have a big impact on culture. [69]

Other forms of engagement included town halls called "Safety Days." These were occasions when senior management visibly acknowledged the hard work contributing to exceptional performance and celebrated workers who had shown outstanding safety performance and could serve as role models. Ultimately it was all about generating a new conversation that would reshape the mindset, behavior, and actions of all stakeholders.

Stakeholder Engagement

Compared to other complex construction projects, Pearl GTL created new engagement benchmarks—by management, supervisors, and workers alike. To illustrate, here is a sample of initiatives:

- Creation of a comfortable living environment to qualify as a "home away from home."

- High-quality food, customized to the local cultures and cuisines of the workers and managers.
- Recreational facilities that catered to all employees' physical and mental needs, including sports pitches, well-being centers, and free Skype for workers to communicate with their families and loved ones.
- A confidential "Aunts and Uncles" program to help workers deal with personal issues and distractions.
- Regular group celebrations.
- You have to stand in the employees' shoes and find creative ways to engage key stakeholder groups in ways relevant to them. For example, with over 50 percent of the workforce stemming from India, the megaproject leaders came up with a pleasant surprise: They invited Kapil Dev, a famous Indian Cricketer, to speak, connect and engage with the workers.

The return on investment made all these efforts pay off—the result: the most successful project safety performance in the history of Shell.

Shell Pearl GTL Success Factors

Project Phase	Critical Success Factors	Beliefs & Biases	Impact
Governance	Authentic commitment to safety and well-being demonstrated through leadership behaviors and role modeling. Adequate funding for creating Incident Free Performance for Shell's single largest project investment to date.	Massive transformations require leadership and resources.	Ambition viewed with credibility.
Business Case	A large-scale project to meet global demand for cleaner energy. One of Shell's landmark international projects. Conversion of natural gas into cleaner-burning products such as diesel, oils for advanced lubricants, naphtha to make plastics, and paraffin for detergents. The detrimental impact of accidents and incidents on people's lives, the success of the project, the brand of the company, and the reputations of	Shell's move to a cleaner economy.	Robust business case.

	stakeholders.		
Final Deliverable	The world's largest Gas-to-Liquids (GTL) plant. Establishment of a strong safety culture. 77 million person-hours without a single Lost Time Injury (LTI).	Culture shapes performance.	World-class asset delivered safely.
Procurement	Safety performance is an essential criterion in the contracting process evaluation.	Encode commitments into contracts.	Supply Chain Commitment ↑
Delivery	Celebrity endorsements. Unyielding focus on the "human angle" of project safety performance: Why should each of us care? Training thousands of supervisors and craftsmen at the Pearl GTL plant on the lessons learned from Shell's previous global experiences. Cascaded engagement of the project supply chain. Initiatives with tangible impact on working conditions (e.g., accommodations, the	Engagement key to success.	Safety Risk↓ Global Reputation↑

	living environment at Pearl Village, physical and mental well-being programs).		

Table 13.1: Shell Pearl GTL's Key Success Factors Along the Flight Path

We have now come to the end of the four mini-cases. We trust you have already discovered some principles and/or practices you can apply to your project. In the next chapter, we bring it all together: We show you a sort of "Ten Commandments" for leading your own project to success. You can use the chapter as a checklist: If something goes awry in your project, match it against these principles and practices to see what's missing.

Radar Alert 14

AS A PROJECT LEADER,

YOU HAVE EXACTLY <u>ONE</u> CHANCE TO SUCCEED:

THE DATASET IS SMALL (N=1).

STEERING YOUR PROJECT

WITH AN EFFECTIVE CHECKLIST

MAKES A DIFFERENCE.

Chapter 14 > Piloting to Success: 10 Anti-Crash Instruments

Dialogue is the basic unit of work in an organization.
Almost all work gets done through dialogue.

— Ram Charan

Our stimulus for writing this book was to leave you with greater power to think differently about your projects, with a greater capacity to conceptualize and realize new ideas. Whether you are kicking off a new initiative, wading through an existing troubled project, fighting fires, or attempting to recover the original plan, you may be wondering: "How can I apply the ideas in this book to the immediate challenges facing me?"

The Power of the Bird's Eye View

The first step in leading projects more effectively is developing a capability to put some distance between you and the project. We call this shift a move from Working **In** the project to working **On** the project.

The Nobel laureate Herbert A. Simon and his psychologist colleagues Michelene Chi, Paul Feltovich, and Robert Glaser discovered that experts and novices see the same problem differently. Beginners often jump to fixing the immediate problem. And not just newcomers: Jumping to the wrong conclusion is something we're all familiar with. The same goes for our frustration on discovering we have wasted heaps of time and energy working on the incorrect problem. Both of these are symptoms of novice thinking. On the other hand, experts tend to spend more time on the front end. They focus on enlarging their understanding of the "bigger picture." A fundamental element needed to see the big picture is the capacity to comprehend and grasp the critical interconnections of seemingly disparate parts of the puzzle. Experts do all this before they develop their action strategies. [70]

10 Checks Along the Flight Path

Check 1: From Working *In* to Working *On* the Project

Moving from "*In*" to "*On*" requires you to step out of the Flight Deck, get off the plane, walk down the steps, cross the tarmac, and take the elevator up to the Air Traffic Control tower (ATC).

Sounds simple, right?

The problem with "weighty" and substantial megaprojects is that they lure project boards and delivery teams into tangled webs of complexity. With deadlines looming, key performance indicators blinking "condition red," and impatient clients or bosses breathing down your neck, it's easy to fall prey to feelings approaching panic.

Indeed, we have both seen many regular, well-adjusted, seasoned executives suffer from deer-in-the-headlights syndrome when a project appears to be heading towards the edge of the cliff. The gravitational pull for Working **In** the project becomes much more prominent than Working **On** the project. The focus narrows, the bigger picture disappears, team members become blinkered and all that's left is tunnel vision.

Central to Working On (not In) the project is the practice of displaying your Project Flight Path (or at least a key segment of it) in front of you. This physical separation between you and the project allows for collaborative conversations between all key parties involved in the various stages of the project life cycle. Teams can co-develop a shared understanding of the complexity of the delivery environment and the issues at play. They can align on the key specs any solution would need to meet.

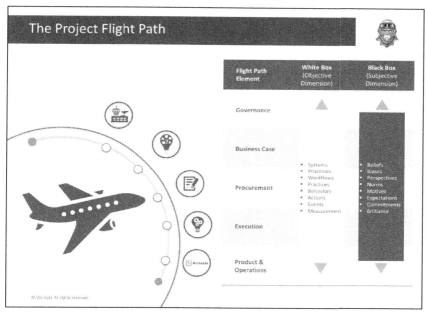

Figure 14.1 The Project Flight Path

Check 2: Facts vs. Fiction

A crucial aspect of Working On, stepping back and taking the bird's eye view, is separating the "Facts" of the project's Flight Path from any "Fiction" that may be present. "Fiction" in this context refers to the world of judgments, opinions, fixed beliefs, and limiting conclusions—accurate or false—held by key stakeholders and project delivery teams. We call all these Fiction since they are ultimately all stories. In our experience, an emotional component is also thrown into the mix, for example, simmering resentments, unexpressed upsets, or repressed anger. What's common to all of these emotions is that they focus on the past. If left unchecked, they drag the past into the future and perpetuate it. When unaddressed, Fiction acts as a palpable invisible force that can severely impede any project's Flight Path.

Let's conduct a little experiment. For one day, write down what people in your project say to you or each other. Then label each statement on your list as either "Fact" or "Fiction."

"Those guys have no idea what they're talking about." "I'm about to explode." "Stop whining. That won't get us anywhere." "I didn't get enough sleep." "The purchasing guys have their heads up their a___s."

Not to spoil it for you, but you will probably find that north of 90 percent of all the sentences is not Fact but Fiction. "I don't have time" is not a Fact; it's a story.

Used in the way described, the Project Flight Path acts as an ***instrument of rationalization,*** allowing everyone to simultaneously see the White Box and Black Box factors driving the megaproject's performance. On a single canvas, you can see both the current results of the project and the system of thinking in the Black Box that drives the outcomes.

Check 3: Psychological Safety

Central to an effective Flight Path is a safe atmosphere for people to tell the truth, express their concerns and reveal their fears without being censored or punished for speaking up. As a leader, the look on your face and the mood you create around you are critical. Working On is not a witch hunt. It's not about finger-pointing or blame games or placing a victim in your crosswire before taking the shot. Remember: It can take courage for people to:

- Surface the "pink elephant in the room."
- Address areas poisoned by mistrust, suspicion, coercion, manipulation, or second-guessing.
- Hold each other to account for promises made and results delivered (or, even more so, not delivered);
- Authentically say "I/we don't know," instead of having to pretend to know the answer when they don't: and
- Openly confront breakdowns in performance, speed, cost, or quality.

In the absence of an atmosphere of straight talk, you can expect to see "false greens": filtering information and reporting only positive data, creating the illusion that the project is doing better than it is.

The costs of covering up gaps in performance can be huge. In the mid-1980s, Xerox executives had a rude awakening when an important new product, the 5046 mid-range copier, failed in the market because of severe reliability problems. Given the company's focus on Leadership Through Quality (LTQ), a quality-assurance process set up by then-CEO David Kearns himself, how could this happen? After investigating, Kearns found that Xerox managers (at all levels) had been all too aware of the problems but had conspired to keep them under the carpet. The cover-up had extended to the highest levels of management; the faulty copier would have never been brought to market but for this code of silence. What particularly distressed Kearns was that Xerox had precisely set up the LTQ process to prevent fiascos of this kind. But the process had failed because "the old culture of our people being afraid to deliver bad news had not yet rinsed from the company."[71] In all fairness, Kearns himself was not entirely innocent; some said he had failed to eliminate multiple management layers and had not taken down the walls between departments. Because of this twin problem, crucial information never reached the top. Whatever the cause, the bottom line is that managers did not make the breakdown public. It cost them their jobs and jeopardized Xerox's reputation and bottom line.

"I was never allowed to present to the board unless things were perfect," said a former Xerox executive. "You could only go in with good news." Whoever gave bad news risked being blamed or booted out: The board would simply kill the messenger. And when directors finally forced managers to confront their poor performance, executives repeatedly blamed short-term factors – from currency fluctuations to trouble in Latin America. By the time then-president (and later CEO) Anne Mulcahy came out and spoke the truth – the company had an "unsustainable business model," she told analysts in 2000 – it was too late; Xerox was already flirting with bankruptcy. (Mulcahy's commitment to transparency and full communication was critical in Xerox's recovery.)

This happened precisely with another case: the colossal £20 billion Crossrail project and its frequent budget blowouts. Mark Wild, the CEO of Crossrail, put it this way: "We were presenting the project as being 94% complete when it was only about 60% done." Fortunately for Wild, all of this had happened before his tenure, and his leadership steered the project onto a more reliable flight path. Open communication was vital to Wild's leadership and success. "Crossrail isn't a business," he said; rather,

it's there to execute a program, so this is a bit of an unusual situation. Through the experiences in my career so far, my belief is that progress only really happens through the people. The technology is secondary. I have worked in some massively complex, safety-critical technology businesses, and the key thing I have learnt is, that its people are the enabler to make it all happen. In general, leadership happens through conversation. I haven't yet met anyone who can refute that. I believe in human beings talking to each other in the spirit of collaboration to achieve a goal. [72]

Check 4: Revealing Beliefs—Turning Around an Airport Project

Turning around a project in deep trouble, with toxic mindsets, deep-seated resentments, mounting mistrust, and active commercial warfare can be quite a challenge. This is precisely the situation that faced a hefty (US$700 million), complex state-of-the-art Asian airport project before they approached us to assist in altering the project's flight path.

With a program running half a year behind schedule and fractured relationships between the client, the delivery Joint Venture, and significant supply chain partners, we asked the top leadership teams to peek into the project's collective Black Box. What the leaders saw astounded them. At least five beliefs and biases dominated the entire Flight Path:

- A fundamental belief that the other parties were the enemy, and "they're out to get us."
- Cosmetic conversations that protected commercial positions and hid severe risks.
- Group despair and resignation about delivering the program on time.
- Promises made with a weak (or no) intention to follow through. And
- A Leadership team that only saw problems and did not recognize or celebrate milestones and accomplishments.

Revealing these beliefs became a pivotal starting point for transforming perspectives that stifled performance across the entire project. Shifting viewpoints was central to driving new behaviors and actions. The ultimate result was that the project was delivered on time, and commercial differences were resolved within a few months of completion. The latter outcome was something nobody believed was possible at the start of the engagement.

Check 5: The Research Phase—From Big Data to Deep Data

Stepping into the control tower allows you an aerial view of the plane (the project). From this vantage point, you can ask powerful "Why," "What," and "How" questions to understand the machinery of beliefs and biases running in the darkness of the Black Box. Doing this allows the leadership team to grasp the project's flight telemetry and obtain essential data on:

- The project's speed, direction, pitch, and progress.
- The areas of uncertainty and how the project's risk profile will change over time.
- The invisible hurdles and impediments lurking below the surface so that the project can respond more effectively; and
- The prevalent mindsets of the various project teams who must tackle the various project issues.

Unpacking the beliefs and biases running in the Black Box is far from trivial. When performance stalled on a USD 1.2 billion (key) component of China's massive high-speed rail program, this intervention revealed the project's DNA and allowed the project leadership team to get the project unstuck.

Of significant note was that this project had a history of revolving doors: project directors coming and going. In short, they were either fired or resigned. The project was on director Number 5 when the project engaged us. Losing the top person on a fast-moving project can be detrimental. Not only do you lose the person, but you also lose the history and institutional memory that went with them, the contextual knowledge and understanding, and the momentum they brought to the project. If you have ever experienced losing a key staff member, you know the impact we are talking about. It can be excruciating. And this project was in pain: It had lost (or removed) four leaders. That was the project's status quo. Add to this calamity a continuous stream of negative publicity and the specter of a media poised and ready to pounce on any wrongdoings. Within this context, people had the following to say about working on the project:

"It's a cauldron of frustration and confusion."

"People are tired and demoralized. The job is relentless. Everything is urgent."

"It has become an ultra-marathon in the desert."

"This job has gone from being a highlight on my CV to asking myself, 'Should I even mention it?'"

Once these conversations were out in the open, the project teams could examine them, identify the valid core of truth in each, and let go of the ones that turned out to be myths. The conversations running in the Black Box held power over the teams only to the extent that they were in the Background and the team had not dragged them to the light.

Check 6. Activate Your Sonar

Applying the Flight Path approach and tools, the leadership team was able, for the first time, to step back from the daily blizzard of activities and see the inner wiring of their project. They went from Working In to Working On. For the project, the impact was analogous to Neil Armstrong's famous quote when he was the first human to step on the moon's surface: "One small step for man, one giant leap for mankind." Commercial and performance issues that had been stuck for months, even years, started to move and gain momentum, even at the late stage when we had been hired to make the impact.

One member of the project leadership team put it this way: "It feels like we have stepped out of the dark and into the light." Another project leader said: "Seeing how all the jigsaw pieces in the Black Box contributed to what is happening makes us think about the project from an entirely new perspective." And that is precisely what had happened: The team had developed a much richer and fuller picture of their Flight Path. The leadership team had now put its hands on the engine and instruments of changing course and turning the project around.

Check 7: Leverage Brilliance Through Conversations

It's not all bleak: The third "B" in the Black Box is Brilliance. Biases and limiting Beliefs can hold the project back; Brilliance does the opposite: The untapped Brilliance lying fallow in your team can ignite quantum leaps in performance. "For every pair of hands," as one of our clients put it, "we get a free brain."

Brilliance lives in the quality and depth of the daily conversations on the project. In fact, the fundamental flaw in project management is simple: People think projects are things. We suggest they are conversations. This is not the truth--it's an empowering interpretation. It's simply a way to look at projects that seems more in tune with how projects actually work. And this perspective puts your hands on the steering wheel of making change happen: Changing conversations that drive everything is a lot easier than changing something that is fundamentally fixed.

So, imagine for a moment that projects are not fixed things but rather networks of conversations. That is one of the core assumptions in this book. If we adopt this assumption, then we say some radically new things.

- Tasks are conversations.
- Processes are conversations.
- Commitments are conversations.

- Agreements (with colleagues, contractors, or clients) are conversations.
- Plans are conversations.
- Performance is a conversation.
- Reports are conversations.
- Monitoring is a conversation.
- Even failures are conversations.
- And so on.
- The bottom line: A project, any project, is a box of conversations.

Check 8: Project Success or (Failure) Lives in the Network of Conversations

Conversations are more than a mere transfer of information or messaging between parties. Conversations allow you to take your project out of "neutral" and move through the gears of performance. Powerful conversations:

- Provide context and meaning
- Build relationships and stakeholder commitment
- Engage and energize individuals, teams, and organizations
- Deepen trust and loyalty
- Inspire and provide the bigger picture
- Generate organization and task clarity
- Mobilize action, coordinate expectations, and identify problems
- Foster innovation, generate solutions and reveal learnings

Through clear and effective conversations, the Guggenheim project leaders ensured the estimated costs were as realistic as possible. They did this by verifying and ensuring the costs were based on completed designs and held the supply chain accountable for their numbers.

Through clear and effective conversations, Shell Pearl GTL built an authentic commitment to people's safety, from the senior leadership to the most remote elements of the delivery chain.

Check 9: Silence Is Not Golden

Conversely, stepping over tricky issues, turning a blind eye, or simply avoiding them is a surefire way to build up risk in your project. You can learn a lot by observing what people say about your projects. Actively look out for:

- People raising problems but not providing solutions.
- Conversations that move around in circles with little forward momentum; and
- Unaddressed thorny issues or taboos, leaving people stuck in their seats and terrified of putting their hand up.

Consider the impact on the Joint Strike Fighter and Sydney Opera House megaprojects if the following questions had been welcome for exploration:

- **Sydney Opera House**: "What challenges do we face turning this idea into reality? What do we need (capabilities, resources, processes, and methods) to avoid wasting valuable taxpayer dollars?" Or, more simply: "What's missing, and what are the barriers?"
- **Joint Strike Fighter**: "Given the history of multi-mission aircraft failures, how exactly will these new technologies overcome the conflicting requirements of the army, navy, and air force?" Or, more simply: "What could go wrong?"

Thinking about a project as a Network of Conversations is a powerful organizing principle for boosting performance. Observing the mood and effectiveness of the conversations and intervening when necessary is often a missing ingredient in leadership effectiveness.

Check 10. Integrity in Action

It is one challenge to lead by example and live a life of integrity as an individual. It is quite another to build or restore integrity for a company or institution. Take the Volkswagen emissions scandal[73] (also known as Dieselgate) that broke in September 2015 when the U.S. Environmental Protection Agency (EPA) issued a notice of violation of the Clean Air Act to German automaker Volkswagen Group. The agency found that the company had intentionally programmed turbocharged direct injection (TDI) diesel engines to only activate their emission controls during laboratory emissions testing. During regulatory testing, this tampering caused the vehicles' NOx output to meet U.S. standards—when in fact, they emitted up to 40 times more NOx in real-world driving. From 2009 to 2015, Volkswagen deployed this programming software in some 500,000 vehicles in the United States and 11 million cars worldwide.

The EPA's charge: Volkswagen had insisted for a year that discrepancies were mere technical glitches. Only after being confronted with the evidence to the contrary did Volkswagen fully acknowledge that it had manipulated the vehicle emission tests.

The first sign that the company might be ready to come clean reportedly came on 21 August 2015 at a conference on green transportation in California, when an unnamed whistleblower approached the director of the EPA's Office of Transportation and Air Quality and surprised him with the informal admission: The company had systematically deceived regulators.

To control the damage, Volkswagen's then-CEO Martin Winterkorn said: "I personally am deeply sorry that we have broken the trust of our customers and the public." Winterkorn, at the helm of Volkswagen since 2008, attributed the wrongdoing to "the terrible mistakes of a few people." But this did little to calm the public outcry. Volkswagen Group of America CEO Michael Horn was more direct than Winterkorn, saying, "We've totally screwed up." Horn added, "Our company was dishonest with the EPA, and the California Air Resources Board, and with all of you."

The fallout was both swift and sweeping. Winterkorn initially resisted calls to step down but finally resigned in late September 2015. Other senior managers, including the head of brand development Heinz-Jakob Neusser, Audi research and development head Ulrich Hackenberg, and Porsche research and development head Wolfgang Hatz, were suspended. Volkswagen's stock price fell by one-third immediately after the news broke.

The automaker took action to save its reputation, but far too little too late. More than half a year after the outbreak of the scandal, it announced plans in April 2016 to spend €16.2 billion (US$18.2 billion) on rectifying the emissions issues and refitting the affected vehicles in a recall campaign.

In April 2017, a US federal judge ordered Volkswagen "to pay a $2.8 billion criminal fine for rigging diesel-powered vehicles to cheat on government emissions tests." In May 2018, Winterkorn was charged in the United States with fraud and conspiracy. Regulators in multiple countries began to investigate Volkswagen.

In an unusual move already in late September 2015, Switzerland had acted more swiftly and decisively than Volkswagen did: The country banned sales of Volkswagen diesel cars altogether. It marked the most severe step taken by a government in reaction to Dieselgate.

And these were only the legal and commercial consequences. The social and human costs would buffet Volkswagen for years to come. A peer-reviewed study published in *Environmental Research Letters* estimated that approximately 59 premature deaths would be caused by the excess pollution produced between 2008 and 2015 by vehicles equipped with the device in the United States, the majority due to particulate pollution (87%) with the remainder due to ozone (13%). The study also found that making these vehicles emissions-compliant by the end of 2016 would avert an additional 130 early deaths.[74] Such is the difference integrity (or the lack thereof) makes in organizations. Corporate culture matters, and here is the irony: On Volkswagen's website, the core value that tops the list is "Integrity: We always strive to do the right thing. Our commitment to the truth is unwavering, both in actions and in words." But do VW managers and workers live by these organizational values in their day-to-day actions or merely pay lip service?

The ultimate measure of success comes down to "Did we deliver on our promises?" Did our actions match our words (promises) to our stakeholders, investors, regulators, key users, partners, and the public? As a project leadership team, did we own the full array of expectations that came with the project? From this perspective, leadership reputation can be summarized as:

LEADERSHIP REPUTATION = BOLD PROMISES MADE X FREQUENCY OF KEEPING THEM

Moreover, the size of your leadership correlates with the size of your promises. In high-visibility megaprojects, you must be aware of *explicit promises* (those you made verbally) and *implicit promises*, which people expect from you as a respectable professional.

If you go to a restaurant and ask for a glass of water, you have an implicit expectation that the glass will be clean—no yucky lipstick marks—and the water will be drinkable. And if you happen to be in China, the water is usually hot, which also brings us to the issue of being sensitive to meeting local cultural norms.

Ultimately, what is behind corporate scandals is simply broken promises. What cost Winterkorn his coveted CEO job in Dieselgate is that he broke his word. But Winterkorn isn't alone. Think of Enron (broken promises regarding the accurate filing of company financials), Uber's sexual harassment (broken promises as an employer), and Apple's 2017 Batterygate (broken promises regarding product safety). The list could go on for a long time. But you see the point: Managers failed to live up to their word.

To be clear. we are not talking about integrity as a moral phenomenon here. We are simply talking about congruence between word and deed. And we say that your performance is directly tied to your integrity. If a bicycle wheel is flawlessly designed, it will run impeccably. Or if a piece of software is flawlessly coded to specs, it will function impeccably, with no bugs. People and projects are no different.

The Cleaner Fuels Project

Broken promises and lack of action were hampering the performance of a USD 500 million Cleaner Fuels Upgrade Project when we were approached to help boost performance. Here is a snippet of the first meeting with the project director:

Vyas: "The program you sent seems thorough."

Philip (not the person's real name): "Yes, it is."

Vyas: "The Excel file with the Risk Matrix is a great piece of work. Looks like the team has spent a lot of time on the visuals and dashboards."

Philip: "Thanks. They have."

Vyas: "So what's the current status? How's the project performing?"

Philip: "Well, that's the point. Performance is poor. Our people have a mindset where they feel they have done their job once the risks have been identified and placed on the register. The gap between the spreadsheet and actual actions to reduce the risks is immense...".

The ultimate purpose of megaprojects is to turn ideas into reality. But statistics show how frequently projects struggle in the action. Plans do not automatically turn themselves into action. The plans and actions in your divisions, workstreams, and projects sites are not the same things but two distinct phenomena. That sounds trivial, but it's surprising how many people confuse and collapse the two. In reality, performance comes down to the hour-by-hour, day-by-day micro-conversations, micro-choices, micro-decisions, and micro-actions. This is what moves the project along the Flight Path. This, in turn, is based on what people see (their perspective), the quality of the dialogues, the working environment (safe or not) and, ultimately, action.

We have repeatedly seen how top managers in the boardroom develop strategic objectives; then, they build a PowerPoint deck, and one board member shows the deck to the middle managers, assuming that this is enough to get buy-in and implement the project. And when people react with silence, the top executive takes that as consent. Nothing could be further from the truth. The chief exec has no idea what's going on in the heads of the middle managers. They might be skeptical, resigned to their fate, or quietly furious that nobody consulted with them in advance. The ability to read and powerfully address the background context is a recurring under-developed capability found in many corporate initiatives.

Now that you have these ten flight instruments under your belt, as it were, we can risk a glimpse into the future. The next chapter explores how machines and artificial intelligence is bound to impact the world of project management.

Radar Alert 15

TECHNOLOGY IS A TOOL, NOT THE MASTER

(AT LEAST NOT YET).

USE IT, BUT DON'T SURRENDER YOUR LEADERSHIP.

Chapter 15 >
Megaprojects and
Intelligent Machines

Any sufficiently advanced technology

is indistinguishable from magic

— **Arthur C. Clarke**

Both of us, Vip and Thomas, came of age in the 1980s, when countless kids and many adults would sit glued to their TV sets, eagerly watching cartoons like "The Flintstones" or "The Jetsons." The two shows were opposites. "The Flintstones" was a spoof based on our stone-age past. "The Jetsons" explored, speculated, and dreamed up how the world could be, and asked, "What if X were possible?" "The Jetsons" captured the imagination and excitement of how people and technology might coexist in the future. The family of four (George and Jane Jetson and their kids Elroy and Judy) lived in a Skypad apartment, high above the ground, in Orbit City. How cool is that! Imagine waking up in the morning, sipping your hot chocolate, and taking in the aerial, panoramic view of the entire metropolis beneath you. It was a life of ease and effort-free living, with few inconveniences. George Jetson's work commitments ("hard labor") amounted to working one hour a day, two days a week. Not sure if you can still call that work-life balance.

In this world, shopping malls floated high up in the sky, and flying cars zipped across in air-lanes. Hologram movies beamed your favorite Hollywood celebrities directly into your living room. Then there were a host of small, push-button labor-saving gadgets that removed those annoying, trivial chores from life. The one we liked the most was the robot contraption that automatically washed, ironed, and folded our clothes. And all this was made possible because a host of advanced technologies had become interwoven into the fabric of society.

When the cartoon was released, most of these ideas were still far-fetched, if not pure fantasy and science fiction. It was an era with rotary phones and phone booths, typewriters, and telex machines, when "Apple" just stood for fruit and "Amazon" for a river; when we still did not have our first computer, let alone the Internet. Fast-forward 40 years, and guess what? Many Jetson ideas are real, alive, and kicking today. People hover above the ground using jetpacks, 3-D printing techniques allow you to build your own tooth implant, a gun, or an entire house. Precision robots move heavy loads. Fridges run their inventories, see which foods and drinks are running low, and order groceries. The military uses real-time digital data to create an augmented battlefield reality to improve split-second decision-making. Companies adopt surveillance drones to track operations, conduct safety checks, and inspect field performance. Smart sensors check your body. Special carpets placed next to each bed in senior homes can now monitor if the elderly guest has stood on the rug by 8:00 am, and if not, automatically alert the nurses. Smartwatches check the real-time health of workers. Algorithms are better at diagnosing illnesses than doctors, better at picking the right hire than HR people or bosses, and better and faster at trading than traders.

We agree that some of these examples are a bit creepy and disturbing. We are not making a value judgment here. Our point is simply that the age of the machine is already here. If you have been in a project office or a field site lately, you've seen how these technologies impact major and mega-projects.

Delivering Projects in a "Jetsons" World

We stand at the cusp of a new era: The impact of Artificial Intelligence and other disruptive technologies on megaprojects is both undeniable and inevitable. Already, we are seeing extensive use of Digital Twins in industries as diverse as automotive, manufacturing, construction, healthcare, and utilities. For those unfamiliar with Digital Twins, think of them as a connector between the real and digital worlds. They work by creating virtual simulations or clones of existing physical assets, hence the term "Digital Twin." Using sensors built into the physical assets, virtual simulators mine and analyze data, showing managers how various elements and processes in the physical device work together. This visibility is critical in helping design, build, operate and maintain physical assets that deliver higher performance at lower costs.

Zoom out from Digital Twins of individual products or assets, and we end up with a Flight Path of entire cities. "Smart Cities" gather big data through various electronic methods (sensors, voice activation, facial recognition, etc.) The data is then used to monitor, fix, and manage the city's assets, resources and services more efficiently. Examples of applications include power plants, water supply systems, schools, sanitation, waste management, hospitals, more efficient public transportation, smoother mobility across the city, and ultimately solving social problems such as early crime detection, prevention, and increasing public awareness of safety.

Zooming further out, several prominent technology companies have invested billions of dollars in the Metaverse. This Digital Twin, on a global level, promises to build a virtual reality that mirrors in every way our reality. It's a clone of the world. The Metaverse is bound to be a greater disruption than the Internet was in the 1990s. It will transform how we live, work, and play. But even without the

Metaverse, megaprojects already benefit from technological advancements:

- IT. makes project tracking easier through real-time project updates;
- IT improves virtual collaboration between decentralized, globally dispersed teams;
- IT improves efficiency and productivity through automation of time-consuming processes, and
- IT enhances effective decision-making through Increased transparency and data visualization.

Fig. 15.1 below shows the considerable variation in technology spending as a percentage of revenues between different industry sectors. Banking and Securities spend on average approximately 7 percent, compared to only 1.51 percent for the construction sector. In addition, there are also considerable variations regarding the benefits expected from these project investments.

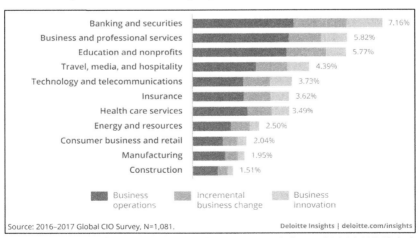

Source: 2016–2017 Global CIO Survey, N=1,081.　Deloitte Insights | deloitte.com/insights

Figure 15.1: Technology Spending by Industry. Source: Deloitte, 2016-2017 Global CIO Survey

A deeper look at construction, the lowest-investing sector, shows how technology investments primarily focus on improving business operations and efficiency. Very little is invested in innovation and top-line growth

To build on these findings, Figure 15.2 below illustrates the potential contributions of disruptive technologies to a project's Flight Path. As a starting point, project boards could deploy Artificial Intelligence (AI) and Machine Learning (ML) in their decision-making. Visualization tools such as Digital Twins can assist in building the business case. The Metaverse can allow for prototyping a virtual version of the final product. AI could also be used for estimates and schedules based on previous similar projects. A Blockchain can enhance effective management of the supply chain through Smart Contracts, which are computer programs (transaction protocols) stored on the Blockchain and programmed to run when predetermined conditions are met. Smart Contracts typically serve to automate the execution of a process so that all participants can be immediately sure of the outcome, without any involvement of an intermediary or any time loss. That's why Blockchains have been called "trust machines." The trust between economic agents that traditionally was provided by banks or notaries or lawyers, or in retail commerce by centralized actors like Amazon, Airbnb, or Uber, can now be provided by the Blockchain and Smart Contracts. Hence Smart Contracts lower transaction costs.

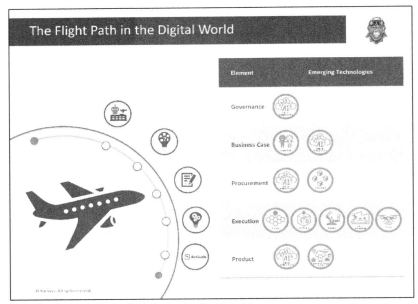

Figure 15.2: The Flight Path in the Digital World

For example, say you have a shipment of rice that you want to transport from the Sichuan province in China to London via the Gulf of Aden, Suez, and Gibraltar. Traditionally you needed several dozen separate bills of lading, a separate contract for each port of call, with lawyers and notaries involved for each contract. Now, you can have a single Blockchain, like a decentralized, transparent, and permanent ledger on which all transactions are recorded. This creates trust. And that trust is not some temporary or psychological jargon. It produces actual savings. According to TradeLens, a company co-founded by shipping giant Maersk and IBM; and dedicated to digitizing the global shipping industry, some USD 18 trillion worth of goods trade worldwide annually. Of these, USD 12 trillion, two-thirds are shipped in containers. The shipping industry has traditionally been a bit slow, to say it nicely. TradeLens says that inefficiencies in global supply chains, primarily facilitated by paper-based logistics, decrease system performance by 15 percent. Using the Blockchain allows shipping companies to manage international trade in the cloud and regain some or all of those 15 percent lost.

Similarly, in the world of megaprojects, Smart Contracts act as a "single source of truth." They enable the project to track every single item and element used in the final product's production. Such transparency can provide the project with up-to-the-minute status regarding the sourcing of items, the locations of production, the status of completion, the identification of risks, the efficiency of quality management, and immediate payment to vendors that make up the supply chain.

In projects where the final deliverables are hard, tangible outcomes (for example, designing and constructing buildings using manufacturing techniques), robotics and built-in sensors during the production can yield real-time data for immediate updating of the project's Gantt Chart. As one of our clients put it, "The impact of Covid is that we can't travel to the production yards. Though not perfect, beaming real-time video from remote sites back to the head office does provide a little more reassurance of the status of key items." This definitely looks Jetsons-like.

The Rise of the Machines

Disruptive technologies like AI, Blockchain, or Metaverse herald new value and wealth creation possibilities for many investors and technologists. But then there is a much larger subset of humanity, people for whom the ascendance of these new machines lives as an existential threat.

Is it possible that Rosie (the name of the Jetsons' housemaid robot) one day gets fed up with serving the morning coffee and turns into a Killer Robot? From one day to the next, the robot's owners, George, Jane, and the kids, become the hunted. We are joking here, but these are genuine concerns for many people. When it comes to our jobs, careers, and employment, the big questions at the back of our minds are, "Will my job become obsolete? Will I be terminated? Worse, will I be unemployable, a little pawn in a world run by a super-intelligence?" These are the ethical, moral, and practical questions living in the Black Box for which solutions have yet to be invented.

In their book, *The Age of AI: And Our Human Future*, authors Henry Kissinger (the former US secretary of state), Eric Schmidt (the ex-Google CEO), and Daniel Huttenlocher (a computer scientist) set out to tackle such dilemmas and describe three possible relationships we can have to disruptive technologies:

1. Confine the technology and its uses where there is a destructive potential for humanity.
2. Partner with the technology, much like a symbiotic relationship, where humans and machines work more effectively side by side. The autopilot on a plane is an example of this.
3. Defer to technology where it's a better performer than we are. For example, computers do much better than us in crunching through large volumes of data and detecting hidden patterns.

Remember: AI Has a Black Box Too

Such distinctions are helpful, but something more sinister, something hidden from our consciousness, also needs to be exposed. Peter Haas, Associate Director of Brown University's Humanity Centered Robotics Initiative, and many others working at the cutting edge of disruptive tech have become increasingly vocal about the detrimental impact of human bias unknowingly (or knowingly) being incorporated into our algorithms.[75] These biases live as lines of code constituting fundamental decision rules.

What type of decisions are we talking about? How about an AI bias that automatically rejects your job application because you happen to be black (or white, or female, or male, take your pick). Remember that a human initially programmed machines. This means the source code reflects that person's subconscious and unconscious beliefs, thought patterns, and blind spots—in short, their bias.

A hidden bias that disproportionately favors one racial group or gender over another in crucial decisions such as hiring individuals is one thing. More chillingly, consider the impact AI bias could have in determining whether someone should be prosecuted or sentenced to prison, and perhaps even the length of their sentence. The bias may not even stem from the source code itself but already lives in the input data fed into the machine at the outset. Worse, few people might even see their own prejudices since programmers or app developers are rarely trained to detect bias. Worse still is the blind obedience to authority that the media and the public grant the output from AI, much like in Stanley Milgram's experiment we highlighted above. No questions asked.

So yes, new technologies have enormous potential, but without solid human oversight, they could do as much harm as good. Keeping technology in perspective and maintaining a healthy skepticism without becoming overly cynical is the hallmark of an intelligent leader. Remember: The systems don't own the Flight Path. You do.

OK, enough Cassandra cries. After this warning, it's time to put the whole Flight Path into practice. That's what the final chapter is about: Action.

CHAPTER 16 > WHAT'S NEXT: PUTTING IT ALL INTO ACTION

The only thing

standing between greatness and me,

is me.

— Woody Allen

You have come a long way in this book (assuming you have read up to this point). Let's do a brief recap of our journey so far. We showed you that megaprojects fail not for technical reasons but for human reasons ranging from mindset to ego to communication to politics and culture. We offered you the Flight Path, a step-by-step turnkey system that will produce results if you apply it correctly. Based on brain science and behavioral economics, we showed you the human fallacies that systematically hamper all "rational actors" and how cognitive failures impact projects. We gave you a series of four mini-case studies of megaprojects: two of which failed while the other two succeeded. The examples were selected so you could learn from both successes and fiascos, distilling the impact of the "Black Box" on projects and deriving best practices to support your flight paths. Now the time has come to use the methodology and get into action. Specifically, there are six things you can do now.

> Step 1: Meet the Authors

You're about to leave the oasis, the safe space of this book, and go into the desert, where you'll face droughts, headwinds, storms, and perhaps even bandits. We want to ensure that you have all the necessary equipment and know when the gear is missing or faulty.

Feel free to connect with us on LinkedIn: https://www.linkedin.com/in/vipvyas/

and

https://www.linkedin.com/in/thomasdzweifel/.

We frequently post on project performance. And we're not going anywhere—at least for now.

> Step 2: Download the Project Flight Path Worksheet

As a free accompaniment to this book, you can download The Flight Path Worksheet by visiting our websites www.vipvyas.com or www.thomaszwiefel.com. The Flight Path Worksheet will provide you with decisive performance conversations crucial for de-risking your project.

> Step 3: Visualize & Shift Your Blackbox

If you're leading a major project and feel that an independent assessment of your Flight Path trajectory and project Blackbox would benefit your project, contact us for a free short consultation.

> Step 4: Sign up for the Project Flight Path Masterclasses

Based on proof of purchase, you will be eligible to attend one of our webinars, where we will further develop the concepts presented in this book.

Would you like to explore how to put the Flight Path to work in your projects straight away? One option is to have your project team and key stakeholders attend a Flight Path Masterclass and (or) in-house workshop. Build alignment among multiple stakeholders. Empower your team to develop project leadership capacities and systematically move from "high potentials" to high performers. Learn from case studies and scenarios faced by project leadership teams to reveal (and fill) performance gaps in your project(s).

> Step 5: High-Performance Industry Roundtables

Join the Gorilla in the Cockpit LinkedIn Group and learn from like-minded professionals and experts. Share your successes and best practices and get valuable input on your challenges.

> Step 6: Become a Thought Leader

Contribute three key insights (or at least one) that you got from reading this book—perhaps something you will start doing, stop doing, or do differently—on the Amazon.com book page:

https://books2read.com/u/mg11G0

Cross-fertilize best practices and pivotal success factors. Post a review and share your learnings (3 bullets suffice). It's a win-win-win: You get to take stock of your accomplishments. Others benefit from your insights. And yes, we get a book review (the more 5-star book reviews, the more Amazon gets the word out—and to make a decisive contribution to the profession, we need as many project managers as possible to re-tool).

> Step 7: Learn Through Action

There is nothing like action; action is the supreme virtue. The technology in this book has a potential billion-dollar value, but only that: potential. None of it will make a shred of difference unless you apply it. Consider for a second that a lot of what we know makes little difference. Why? Because we fail to translate insight into reality. It's called the Knowing-Doing gap. So go out there, work with your team, have the conversations, use the practices, fail, correct (maybe fall again) and succeed. As Winston Churchill said: "Success is going from failure to failure without loss of enthusiasm"). We hesitate to tell people what to do, but in this case, we agree with the Nike slogan: "Just do it!" Get away from your desk, away from the office, and into the field. You get the message. Close the book—you can take it with you as a backup. Take the actions and deliver the future. Onward to victory ;-)

Epilogue: Better Data. Better Models. Better Conversations. Better Actions

An object at rest stays at rest
and an object in motion stays in motion
with the same speed and in the same direction
unless acted upon by an unbalanced force.

— Isaac Newton's First Law

In the prologue and throughout the book, we used the metaphor of flying a plane from point A to point B to systematically illustrate the challenges we face when leading a major project.

Whether you are driving a corporate change initiative or a project designed to alter the fabric of society, chances are you will have to deal with storm systems: problems, issues, and challenges coming your way, whether they come in the form of angry stakeholders, dissatisfied customers, controversial media coverage, unreliable sub-contractors, upset employees or intrusive regulators.

By their very nature, megaprojects (and mega change initiatives) are complex organizations. They are political arenas with a mix of multiple stakeholder interests. For the politicians, the project means one thing, quite another thing for the architect, and yet another for the client. Not to speak of designers, contractors, vendors, and end-users: Each has a different, unique perspective and agenda. It's like Akiro Kurosawa's epic film "Rashomon" where four people recount their version of a murder that happened in the forest. The bandit, the samurai, the wife, and the woodcutter each offer testimonies that are self-serving and contradict each other.[76] The same goes for megaprojects. The project is not a monolithic, fixed phenomenon but an idea that lives very differently in the minds of each stakeholder group. We call it a set of conversations. And the more these conversations operate in the background, invisibly, the more power they have over your results.

We have designed this book to give you a leadership advantage. We believe that if you can picture the context shaping the "project story" from multiple perspectives (the politicians, the owners, the financiers, the architects, the regulators, the procurement people, the delivery teams, and the end-users or beneficiaries), you will have a handle on making any project work. Visualizing the project and displaying the data becomes critical to how we lead and direct projects. The big, fuzzy, hairy, awkward, and complex issues are rarely portrayed in the technical and quantitative data presented at project review meetings.

Our second objective in writing the book was to enable you to unpack in real-time the spectrum of issues in front of you and take actions that help move the performance needle in the right direction. We want you to be able to make a surgical, high-leverage intervention to put the project on track (sorry for the mixed metaphors in that sentence). Major projects require leaders to manage many variables in live dynamic environments successfully. Some problems may be straightforward: You can solve them in a few hours or even minutes. Others will be tangled and messy, involving multiple stakeholder groups. These challenges are the ones that leave your team members scratching their heads and struggling for solutions. They will require you to get below the surface and find the real issues underpinning and driving the symptom issue.

We showed the effects mental biases have on the project's flight path. We unpacked the tremendous impact of bias on both performance and outcomes of four selected megaprojects, for example, in the case of the Sydney Opera House and the F-35 Joint Strike Fighter. But we also highlighted the positive impact Frank Gehry's intentional focus on delivering complete designs, working with robust cost estimates, and holding the delivery system to account had on the success of the Guggenheim in Bilbao. Likewise, Shell's Pearl GTL project illustrated the power of leadership and authenticity in winning over 50,000 employees to get behind a game worth playing: the importance of people returning home from hazardous work, safe and well, every day.

Tapping into the Background and revealing the underlying conversations that inspire and motivate people to give their best, something they care about deeply, something that has them "jump out of bed" each morning, is an asset that has gone largely under-utilized by change and project leaders.

The Flight Path is a tool we have designed for you to step into the Air Traffic Controller's room and *check* your plane and your itinerary before you step into the flight deck and head off to the runway. We all know that fixing your plane mid-air is not a good idea. We wish you a successful flight, and may you arrive safely at your destination—on time and on budget.

FURTHER READING

Drucker, Peter. 1999. *"Managing Oneself,"* *Harvard Business Review*, March-April 1999. 65-74.

Eagleman, David. 2011. *Incognito: The Secret Lives of the Brain.* Edinburgh: Canongate.

Groysberg, Boris and Michael Slind. 2012. *"Leadership Is a Conversation,"* *Harvard Business Review*, June.

Kahneman, Daniel. 2011. *Thinking, Fast and Slow.* New York: Farrar, Straus and Giroux.

Kahneman, Daniel, Olivier Sibony and Cass Sunstain. 2021. *Noise: A Flaw in Human Judgment.* New York: Little, Brown.

Kahneman, Daniel, Andrew M. Rosenfield, Linnea Gandhi and Tom Blaser. 2016. "Noise: How to Overcome the High, Hidden Costs of Inconsistent Decision-Making," Harvard Business Review, October.

Searle, John. 1969. *Speech Acts: An Essay in the Philosophy of Language.* Cambridge: Cambridge University Press.

Snowden, David J. and Mary E. Boone. 2007. *"A Leader's Framework for Decision-Making,"* *Harvard Business Review*, November.

Sull, Donald N. and Charles Spinosa. 2007. *"Promise-Based Management,"* *Harvard Business Review*, April.

Thaler, Richard and Cass Sunstein. 2012. *Nudge.* New York: Penguin.

Vyas, Vip and Gilles Hilary. 2016. *"Does Your Organisation Run on Fear?"* *INSEAD Knowledge*, 11 February 2016.

Vyas, Vip and Stuart Doughty. 2017. *"Hidden Neural Patterns Could be Limiting Your Performance,"* *INSEAD Knowledge*, 30 January.

Vyas, Vip and Diego Nannicini. 2017. *"Overcoming the Misery of Mega-Project Delivery,"* INSEAD *Knowledge*, 7 August.

Vyas, Vip and Diego Nannicini. 2018. *"Aim for Transformation, Not Change."* INSEAD *Knowledge*, 8 August.

Vyas, Vip and Diego Nannicini. 2020. *"Is Your Innovation Process a Corporate Illusion?"* INSEAD *Knowledge*, 24 January.

Vyas, Vip. 2020. *"Leadership at a Distance,"* Project APM Quarterly, *Project APM Quarterly*, Issue 303, Summer 2020.

Vyas, Vip. 2021. *"Mastering Paradoxes as a Strength,"* Duke Corporate Education*, March.

Winograd, Terry and Fernando Flores. 1986. *Understanding Computers and Cognition: A New Foundation for Design.* Boston: Addison-Wesley.

Zweifel, Thomas D. 2003. *Communicate or Die: Getting Results Through Speaking and Listening.* New York: SelectBooks.

Zweifel, Thomas D. 2013. *Culture Clash 2.0: Managing the Global High-Performance Team.* New York: SelectBooks.

Zweifel, Thomas D. 2002. *Democratic Deficit? Institutions and Regulation in the European Union, Switzerland and the United States.* Lanham MD: Rowman & Littlefield.

Zweifel, Thomas D. 2005. *International Organizations and Democracy: Accountability, Politics, Power. Boulder* CO: Lynne Rienner Publishers.

Zweifel, Thomas D. 2019. *iCoach: Coach Hat—The Simple Little Formula for Freeing Yourself, Boosting People Power and Changing the World.* New York: iHorizon.

Zweifel, Thomas D. 2019. *Leadership in 100 Days: Your Systematic Self-Coaching Roadmap to Power and Impact—and Your Future.* New York: iHorizon.

Zweifel, Thomas D. and Edward J. Borey. 2014. *Strategy-In-Action: Marrying Planning, People and Performance.* New York: iHorizon.

Zweifel, Thomas D. and Aaron L. Raskin. 2008. *The Rabbi & the CEO: The Ten Commandments for 21st-Century Managers.* New York: SelectBooks.

THE AUTHORS

Vip Vyas is a Socratic advisor to Boards, CEOs, and CXOs helping them catapult their organization and projects to new performance frontiers. He is an enthusiast of breaking up limiting belief systems and creating unique strategic advantages for turning big ideas into reality.

He loves projects simply because they are the vehicle for inventing and shaping the future.

As the originator of the Project Flight Path model, Vip (and his team and suitcase) have traveled across dozens of countries (35 at the last count), working with executive teams to unlock their corporate Black Boxes and achieve results beyond the predictable,

His journey has taken him to work in the mangrove swamps of the Niger Delta (where machine guns and kidnappings are rife), to the scorching deserts of Qatar and the Middle East, to the blustery gales and turbulence of the North Sea, and the meteoric rise of China and the wider South-East Asia.

His first book, *Gorilla in the Cockpit,* co-authored with Thomas, is the product of over a quarter of a century's investment in tackling and turning around highly complex projects across wide-ranging sectors such as Banking, FMCG, Government, Digital Technology, Oil & Gas, Petrochemicals, Power, Mining, Renewables, and large-scale civil infrastructure, with a total aggregate value exceeding USD 70+ billion.

Of all the global megaprojects Vip has been involved in, his most confronting one was on a micro-scale. That of performing live on stage, in a band, in front of 5000 people, at the South Bank next to the River Thames, London, England. His approach to overcoming his gorillas (a flurry of self-defeating thoughts, total fear, and sheer panic) as he stood in front of an eager crowd, waiting to be entertained, has been instrumental in shaping his perspective on leading change in multi-dimensional settings.

British born of Indian descent, Vip now lives in Hong Kong with his young family. He has studied strategy and execution at the Harvard Business School, been a visiting consultant at the London Business School, and an Executive Consultant at the Säid Business School, University of Oxford, for the Hong Kong Government's Major Projects Leadership Programme ("MPLP"). His thought leadership has been featured in many top periodicals, including Forbes, INSEAD Knowledge, and Duke Corporate Education's Dialogue Review.

Connect with Vip if you are interested in creating step-changes in delivering capital projects more effectively, participating in upcoming online classes, viewing insightful podcasts, and valuable thought-leadership for tackling complex change: https://www.vipvyas.com

Dr. Thomas D. Zweifel (pronounced like "Eiffel Tower") is a strategy & performance expert, board member and sparring partner for CEOs and CXOs of Fortune 500 companies. The ex-CEO of Swiss Consulting Group, named "Fast Company" by *Fast Company* magazine, has coached clients on four continents since 1984 to impact project performance, open strategic frontiers, meet business imperatives and seize growth opportunities. In the aggregate, the projects Thomas worked on yielded annual revenue growth of $9+ billion.

(It wasn't always like this. Thomas started out as a ski instructor in Davos, then worked as an actor, construction worker, director, fundraiser and manager before becoming CEO.)

Thomas has worked on projects at ABB, Airbus, Banana Republic, Citibank, ConocoPhillips, Credit Suisse, Danone, Dell, Deutsche Bank, DHL, Faurecia, Fiat, GE, GM, Goldman Sachs, Google, J&J, JPMorgan Chase, Medtronic, Nestlé, Novartis, P&G, Prudential, Roche, Sanofi, Siemens, Starbucks, Swiss Re, UBS, Unilever and Zurich.

Clients in other sectors include the Kazakhstan prime minister and cabinet, various Swiss government agencies, the UN Development Programme, the US State Department, the US Air Force Academy, and the US Military Academy at West Point.

An authority on integrating planning, people and performance, Thomas helps clients ask the right questions, confront taboos, build strategy alignment, and boost productivity. Ultimately his specialty is unleashing the human spirit in organizations—without unnecessary blah-blah, impractical training programs, or false dependencies on high-priced consultants.

Thomas serves as a board member for, among others, Paramount Business Jets, the KH World Executive, the JAFI Board of Governors.

From 2000 to 2020, Thomas taught leadership to 2,500+ students as an adjunct professor at Columbia University and guest professor at HSG (St. Gallen University). He is often featured in the media, including ABC, Bloomberg, CNN, Swiss National TV, *Fast Company* and *Financial Times*.

Thomas is the award-winning author of nine books on strategy and co-leadership, including *Communicate or Die*; *Culture Clash 2.0*; *iCoach*; *Leadership in 100 Days*; *The Rabbi and the CEO* (with Aaron L. Raskin), a National Jewish Book Award finalist; and *Strategy-In-Action* (with Edward J. Borey), a Readers Favorite Silver Award winner.

Born in Paris, Thomas holds a Ph.D. in International Political Economy from New York University. In 1996 he realized his dream of breaking three hours in the New York City Marathon, and in 1997 was recognized as the "fastest CEO in the New York City Marathon" by the *Wall Street Journal*. Today he is mostly running late... He lives in Zurich with his wife and two daughters—plus a young dog named Motek, the only other man in the house. (This is not a complaint).

Connect to Thomas and get free leadership tools, strategies and trainings: https://www.leaders-academy.online

INDEX

A

B

F

G

I

J

K

R

S

T

U

W

Notes

[1] Alexander Budzier & Harvey Maylor (2021). Working Paper: *"Projects: A US $20 Trillion, World-Scale Problem"*. Said Business School, University of Oxford.

[2] Not Inflation adjusted

[3] Alison Beard & Antonio Nieto-Rodriguez (2021). *"The Future of Work is Projects - So You've Got to Get Them Right"* HBR IdeaCast/ Episode 827. https://hbr.org/podcast/2021/11/the-future-of-work-is-projects-so-youve-got-to-get-them-right

[4] Nicklas Garemo, Stefan Matzinger and Robert Palter, 2015, "Megaprojects: The good, the bad, and the better," McKinsey, July.

[5] John R. Searle. 1983. Intentionality: An Essay in the Philosophy of Mind. Cambridge: Cambridge University Press.

[6] Aircraft Accident Investigation Bureau, Ministry of Transport, Ethiopia, Aircraft Accident Investigation Preliminary Report, 10 March 2019. https://leehamnews.com/wp-content/uploads/2019/04/Preliminary-Report-B737-800MAX-ET-AVJ.pdf

[7] "Boeing 737 Max: What went wrong? BBC News, 5 April 2019.

[8] Tyler Durden, "The Best Analysis Of What Really Happened To The Boeing 737 Max From A Pilot & Software Engineer," *ZeroHedge*, https://www.zerohedge.com/news/2019-03-17/best-analysis-what-really-happened-boeing-737-max-pilot-software-engineer

[9] "Stumbles by Boeing C.E.O. deepen a crisis," *New York Times*, International Edition, 24 December 2019, 8-9.

[10] "Beyond pilot trash talk, 737 MAX documents reveal how intensely Boeing focused on cost," Seattle Times, 10 January 2020.

[11] "Beyond pilot trash talk, 737 MAX documents reveal how intensely Boeing focused on cost," Seattle Times, 10 January 2020.

[12] "The Whistle-Blowers at Boeing," The Daily podcast, New York Times, https://podcasts.apple.com/de/podcast/the-daily/id1200361736?l=en&i=1000436137931

[13] Bent Flyvbjerg, What You Should Know About Megaprojects and Why: An Overview, *Project Management Journal* 45:2, February 2014.

[14] https://www.nytimes.com/2019/08/21/magazine/f35-joint-strike-fighter-program.html

15 https://www.bbc.com/news/uk-england-london-47787367

16 https://www.couriermail.com.au/news/why-sydneys-opera-house-was-the-worlds-biggest-planning-disaster/news-story/9a596cab579a3b96bba516f425b3f1a6

17 https://www.kuppingercole.com/blog/hughes/distributed-ledger-use-cases-in-the-financial-industry

18 https://en.wikipedia.org/wiki/Streetlight_effect

19 Michelle Symonds, "15 Causes of Project Failure," *Project Smart*, 13 June 2011.

20 Robert Kennedy, interview, quoted by Ronald Steel, *New York Review of Books*, 13 March 1969, 22.

21 Graham Allison and Philip Zelikow, 1999, *Essence of Decision: Explaining the Cuban Missile Crisis*, Reading MA: Longman, 327-329.

22 Theodore C. Sorensen, 1965, *Kennedy*, New York: Harper & Row, 675.

23 Vip Vyas, Summer 2020. "Leadership at a Distance," Project. Association for Project Management.

24 Nassim Nicholas Taleb, 2007. *The Black Swan: The Impact of the Highly Improbable*. London: Penguin.

25 David Barstow, David Rohde and Stephanie Saul, "Deepwater Horizon's Final Hours," *The New York Times*, 25 December 2010.

26 "Deepwater Horizon's Final Hours," *The New York Times*, 26 December 2010.

27 Deepwater Horizon Accident Investigation Report, 8 September 2010. https://www.bp.com/content/dam/bp/business-sites/en/global/corporate/pdfs/sustainability/issue-briefings/deepwater-horizon-accident-investigation-report.pdf

28 *ibid.*

29 *ibid.*

30 *ibid.*

31 Deepwater Horizon Study Group, "Final Report on the Investigation of the Macondo Well Blowout," 1 March 2011, 5.

32 Deepwater Horizon Study Group, "Final Report on the Investigation of the Macondo Well Blowout," 1 March 2011, 9. http://large.stanford.edu/courses/2011/ph240/mina1/docs/DHSGFinalReport-March2011-tag.pdf

33 Rachel Maddow, 2013, *Drift: The Unmooring of American Military Power*, New York: Crown.

34 Lise Arena and Eamonn Molloy, 2010, "The Governance Paradox in Megaprojects," *Entretiens Jacques Cartier*, Lyon, France. https://halshs.archives-

ouvertes.fr/halshs-00721622/document. See also Ronald H. Coase, 1937, The Nature of the Firm," *Economica,* Blackwell Publishing, **4** (16): 386–405; and Oliver E. Williamson, 1981. "The Economics of Organization: The Transaction Cost Approach". *American Journal of Sociology.* 87:3, 548–577.

[35] Alberto Megias, Juan Francisco Navas, Dafina Petrova, Antonio Candido, Antonio Maldonado, Rocio Garcia-Retamero, and Andres Catena. 2015. "Neural mechanisms underlying urgent and evaluative behaviors: An fMRI study on the interaction of automatic and controlled processes." *Human Brain Mapping* 36, 2853-2864

[36] Daniel J. Levitin, 2015, "Why It's So Hard to Pay Attention, Explained by Science," Fast Company, 23 September. https://www.fastcompany.com/3051417/why-its-so-hard-to-pay-attention-explained-by-science

[37] PWC - https://www.youtube.com/watch?v=BFcjfqmVah8

[38] https://neurosciencenews.com/brain-size-depression-anxiety-16769/

[39] "Americans are dangerously overconfident in their driving skills--but they're about to get a harsh reality check, *Business Insider,* 25 January 2018.

[40] Bent Flyvbjerg. 2008. Curbing Optimism Bias and Strategic Misrepresentation in Planning: Reference Class Forecasting in Practice. *European Planning Studies* 16: 1, January, 3-21.

[41] Daniel Kahneman and Amos Tversky. 1979. Prospect theory: An analysis of decisions under risk, *Econometrica* 47, 313–327. Kahneman, D. & Lovallo, D. 1993. Timid choices and bold forecasts: A cognitive perspective on risk taking, *Management Science* 39, 17–31.

[42] Rachel E. Frieder, Chad H. van Iddekinge and Patrick H. Raymark, "How quickly do interviewers reach decisions? An examination of interviewers' decision-making time across applicants." *Journal of Occupational and Organizational Psychology*, 11 April 2015.

[43] Richard Gregory (1998, p. 5): ". . . a major contribution of stored knowledge to perception is consistent with the recently discovered richness of downgoing pathways in brain anatomy. Some 80% of fibres to the lateral geniculate nucleus relay station come downwards from the cortex, and only about 20% from the retinas. [See Sillito, A. 1995. "Chemical Soup: Where and How Drugs May Influence Visual Perception", in The Artful Eye. Oxford: Oxford University Press, pp. 291-306.]"

[44] Jeff Hawkins and Sandra Blakeslee. 2004. *On Intelligence.* New York: Henry Holt. 6, 107-116.

[45] Stanley Milgram. 1963. "Behavioral Study of Obedience." *Journal of Abnormal and Social Psychology, 67*(4), 371–378.

[46] Jim Collins, 2001, *Good to Great*. New York: HarperBusiness, 130-133.

[47]

http://www.leeds.ac.uk/news/article/397/sheep_in_human_clothing_scientists_r
eveal_our_flock_
mentality

[48] Herd mentality: Are we programmed to make bad decisions? -- ScienceDaily

[49] Opinion | The Rise of the New Groupthink - The New York Times (nytimes.com)

[50] Daniel Kahneman, 2013, *Thinking, Fast and Slow*, New York: Farrar Straus & Giroux, 283-284.

[51] Adapted from Prospect theory - Wikipedia

[52] Marco Iacoboni. 2005. "Grasping the Intentions of Others with One's Own Mirror Neuron System". *PLOS Biology.* **3** (3). 22 February.

[53] This story is largely based on
https://en.wikipedia.org/wiki/Sydney_Opera_House

[54] Biography - John Joseph (Joe) Cahill - Australian Dictionary of Biography (anu.edu.au)

[55] "The F-35 Joint Strike Fighter, the costliest weapon system in US military history, now faces pushback in Congress," *Hartford Courant*, 1 June 2021.

[56] https://www.epi.org/blog/u-s-trade-deficit-hits-record-high-in-2020-biden-
administration-must-prioritize-
rebuilding-domestic-manufacturing/

[57] Sean McFate, 2021, "The F-35 tells everything that's broken in the Pentagon," *The Hill*, 11 March.

[58] "HASC Chair Slams F-35, 500-Ship Fleet; Highlights Cyber," *Breaking Defense*, 5 March 2021.

[59] F-35 pilot responds to claims that the jet is a failure - Sandboxx

[60] Marijke van Putten, Marcel Zeelenberg and Eric van Dijk. 2010. "Who Throws Good Money After Bad? Action vs. State Orientation Moderates the Sunk Cost Fallacy," *Judgment and Decision Making* 5:1, 33-36.

[61] Rowan Moore, "The Bilbao Effect: How Frank Gehry's Guggenheim started a global craze," *The Guardian*, 1 October 2017.

[62] Bent Flyvbjerg, 2005, "Design by Deception, The Politics of Megaproject Approval", *Harvard Design Magazine*, Spring/Summer.

[63] John R. Searle, 1992. The Rediscovery of the Mind: Cambridge MA: The MIT Press. Chapter 8.

[64] Lee Ross, 1977 The Intuitive Psychologist and His Shortcomings: Distortions in the Attribution Process," *Advances in Experimental Social Psychology* 10, 173-220.

[65] "Lee Ross, Expert in Why We Misunderstand Each Other, Dies at 78," *New York Times*, 16 June 2021.

[66] Jeffrey Pfeffer and Robert I. Sutton, 2006, *Hard Facts, Dangerous Half-Truths, and Total Nonsense: Profiting from Evidence-Based Management*, Boston: Harvard Business Press, 7.

[67] *ibid.*

[68] Julian Johnson, Organization Effectiveness Consultant, Qatar Shell, Interview: https://jmj.com/testimonials/organisational-culture-on-the-shell-pearl-gtl-project/

[69] A propos small changes causing big culture shifts: In the 1980s, before Ciba-Geigy's merger with Sandoz to form Novartis, then Ciba-Geigy chairman Alex Krauer came up with a simple intervention in service of sustainability. The problem was that many managers and workers came to the office on weekends and failed to re-charge their batteries or spend quality time with their families. This in turn had a negative domino effect on their morale, burnouts etc. What did Krauer do? He simply turned the heat off on weekends. This tiny change not only saved heating costs; it also promoted a healthier work/life balance and overall sustainability.

[70] D.P. Simon and H.A. Simon, 1978, Individual Differences in Solving Physics Problems, in R.S. Sigler (Ed.), *Children's Thinking: What Develops?* Hillsdale NJ: Erlbaum, 325-342. M.T.H. Chi, P.J. Feltovich, R. Glaser, 1981, Categorization and Representation of Physics Problems by Experts and Novices, *Cognitive Science* 5:2. 121-152.

[71] David Kearns and David Nadler. 1992. *Profits in the Dark.* New York: Harper Business Books. 249-250.

[72] Linda Walmsley, 2019, "Inspiring Leaders - Mark Wild - CEO Crossrail," https://www.walmsleywilkinson.com/news-and-insights/interviews/inspiring-leaders-mark-wild-ceo-cross-rail/

[73] This account is largely based on Wikipedia: https://en.wikipedia.org/wiki/Volkswagen_emissions_scandal

[74] Steven R.H. Barrett et al., Raymond L. Speth, Sebastian D. Eastman, Irene C. Dedoussi, Akshay Ashok, Robert Malina and David W. Keith, "Impact of the Volkswagen Emissions Control Defeat Device on US Public Health," *Environmental Research Letters* 10, 2015.

[75] Peter Haas. 2017. "The Real Reason to be Afraid of Artificial Intelligence" *TEDX Dirigo. https://www.youtube.com/watch?v=TRzBk_KuIaM*

[76] Rashomon, https://en.wikipedia.org/wiki/Rashomon

Printed in Great Britain
by Amazon

19977090R00139